EPA CLEANUP APPROACHES TO CONTAMINATED AND HAZARDOUS WASTE SITES

THE NATIONAL PRIORITIES LIST AND THE SUPERFUND ALTERNATIVE

WASTE AND WASTE MANAGEMENT

Additional books in this series can be found on Nova's website
under the Series tab.

Additional e-books in this series can be found on Nova's website
under the eBooks tab.

WASTE AND WASTE MANAGEMENT

EPA CLEANUP APPROACHES TO CONTAMINATED AND HAZARDOUS WASTE SITES

THE NATIONAL PRIORITIES LIST AND THE SUPERFUND ALTERNATIVE

ALLAN R. PAYNE
EDITOR

New York

NOTICE TO THE READER

Library of Congress Cataloging-in-Publication Data

ISBN: 978-1-63485-260-9

Published by Nova Science Publishers, Inc. † New York

CONTENTS

Preface   vii

Chapter 1   Superfund: Trends in Federal Funding and Cleanup
of EPA's Nonfederal National Priorities List Sites   1
United States Government Accountability Office

Chapter 2   Superfund: EPA Should Take Steps to Improve Its
Management of Alternatives to Placing Sites on
the National Priorities List   45
United States Government Accountability Office

Index   93

PREFACE

Under the Superfund program, the Environmental Protection Agency (EPA) places some of the most seriously contaminated sites on the National Priorities List (NPL). At the end of fiscal year 2013, nonfederal sites made up about 90 percent of these sites. At these sites, EPA undertakes remedial action projects to permanently and significantly reduce contamination. Remedial action projects can take a considerable amount of time and money, depending on the nature of the contamination and other site-specific factors. This book examines, for fiscal years 1999 through 2013, the trends in the annual federal appropriations to the Superfund program and EPA expenditures for remedial cleanup activities at nonfederal sites on the NPL; and the number of nonfederal sites on the NPL, the number of remedial action project completions, and the number of construction completions at nonfederal NPL sites. Furthermore, the book examines how EPA addresses the cleanup of sites it has identified as eligible for the NPL; how the processes for implementing the Superfund Alternative (SA) and NPL approaches compare; and how SA agreement sites compare with similar NPL sites in completing the cleanup process.

In: EPA Cleanup Approaches ...
Editor: Allan R. Payne

ISBN: 978-1-63485-260-9
© 2016 Nova Science Publishers, Inc.

Chapter 1

SUPERFUND: TRENDS IN FEDERAL FUNDING AND CLEANUP OF EPA'S NONFEDERAL NATIONAL PRIORITIES LIST SITES[*]

United States Government Accountability Office

WHY GAO DID THIS STUDY

Under the Superfund program, EPA places some of the most seriously contaminated sites on the National Priorities List (NPL). At the end of fiscal year 2013, nonfederal sites made up about 90 percent of these sites. At these sites, EPA undertakes remedial action projects to permanently and significantly reduce contamination. Remedial action projects can take a considerable amount of time and money, depending on the nature of the contamination and other site-specific factors. In GAO's 2010 report on cleanup at nonfederal NPL sites, GAO found that EPA's Superfund program appropriations were generally declining, and limited funding had delayed remedial cleanup activities at some of these sites.

GAO was asked to review the status of the cleanup of nonfederal NPL sites. This report examines, for fiscal years 1999 through 2013, the trends in (1) the annual federal appropriations to the Superfund program and EPA expenditures for remedial cleanup activities at nonfederal sites on the NPL;

---

[*] This is an edited, reformatted and augmented version of a United States Government Accountability Office, Publication No. GAO-15-812, dated September 2015.

and (2) the number of nonfederal sites on the NPL, the number of remedial action project completions, and the number of construction completions at nonfederal NPL sites. GAO analyzed Superfund program and expenditure data from fiscal years 1999 through 2013 (most recent year with complete data available), reviewed EPA documents, and interviewed EPA officials.

WHAT GAO RECOMMENDS

GAO is not making any recommendations in this report. EPA agreed with GAO's findings.

WHAT GAO FOUND

Annual federal appropriations to the Environmental Protection Agency's (EPA) Superfund program generally declined from about $2 billion to about $1.1 billion in constant 2013 dollars from fiscal years 1999 through 2013. EPA expenditures—from these federal appropriations—of site-specific cleanup funds on remedial cleanup activities at nonfederal National Priorities List (NPL) sites declined from about $0.7 billion to about $0.4 billion during the same time period. Remedial cleanup activities include remedial investigations, feasibility studies, and remedial action projects (actions taken to clean up a site). EPA spent the largest amount of cleanup funds in Region 2, which accounted for about 32 percent of cleanup funds spent at nonfederal NPL sites during this 15-year period. The majority of cleanup funds was spent in seven states, with the most funds spent in New Jersey—over $2.0 billion in constant 2013 dollars, or more than 25 percent of cleanup funds.

From fiscal years 1999 through 2013, the total number of nonfederal sites on the NPL annually remained relatively constant, while the number of remedial action project completions and construction completions generally declined. Remedial action project completions generally occur when the physical work is finished and the cleanup objectives of the remedial action project are achieved. Construction completion occurs when all physical construction at a site is complete, all immediate threats have been addressed, and all long-term threats are under control. Multiple remedial action projects may need to be completed before a site reaches construction completion. The total number of nonfederal sites on the NPL increased from 1,054 in fiscal year 1999 to 1,158 in fiscal year 2013, and averaged about 1,100 annually. The number of remedial action project completions at nonfederal NPL sites generally declined by about 37 percent during the 15-year period. Similarly, the number of construction completions at nonfederal NPL sites generally declined by about 84 percent during the same period. The figure below shows the number of completions during this period.

ABBREVIATIONS

CERCLA	Comprehensive Environmental Response, Compensation, and Liability Act
CERCLIS	Comprehensive Environmental Response, Compensation, and Liability Information System
EPA	Environmental Protection Agency
ICTS	Institutional Controls Tracking System
NPL	National Priorities List
Panel	National Risk-Based Priority Panel
PRP	potentially responsible party
Recovery Act	American Recovery and Reinvestment Act of 2009
ROD	record of decision
SDMS	Superfund Document Management System
SEMS	Superfund Enterprise Management System
Trust Fund	Hazardous Substance Superfund Trust Fund

* * *

September 25, 2015

The Honorable Barbara Boxer
Ranking Member
Committee on Environment and Public Works
United States Senate

The Honorable Cory A. Booker
United States Senate

The Comprehensive Environmental Response, Compensation, and Liability Act (CERCLA) of 1980 established the Superfund program to protect human health and the environment from the effects of hazardous substances.[1] The Environmental Protection Agency (EPA) is the principal agency responsible for administering the Superfund program. According to EPA officials and agency documents, Superfund protects the American public by cleaning up hazardous waste, contaminated sites, or releases that pose an imminent or long-term risk of exposure and harm to human health and the environment. EPA places some of the most seriously contaminated sites on the National Priorities List (NPL), and cleanups of these sites are often expensive and lengthy. At the end of fiscal year 2013, there were 1,315 sites on the NPL—1,158 nonfederal sites (about 90 percent) and 157 federal facilities.[2] Over a 15-year period from fiscal years 1999 through 2013, the Superfund program received almost $23 billion in federal appropriations in constant 2013 dollars, according to our analysis of federal appropriations data.[3]

Some of the contaminants present at NPL sites have included polychlorinated biphenyls,[4] lead, and arsenic. According to EPA documents, the precise human health effect of many chemical mixtures at NPL sites is uncertain. However, hazardous substances found at Superfund sites have been linked to a variety of human health problems, such as birth defects, cancer, changes in neurobehavioral functions, and infertility.

Two basic types of cleanups are conducted under the Superfund program: (1) remedial actions and (2) removal actions. Remedial actions are generally long-term cleanups—consisting of one or more remedial action projects—that aim to permanently and significantly reduce contamination and which can take a considerable amount of time and money, depending on the nature of the contamination and other site-specific factors. A remedial action project is generally the physical work undertaken to address contamination at a site (e.g., sediment dredging or construction of a landfill cap). Remedial action project completions generally occur when the physical work is finished and the cleanup objectives of the remedial action project are achieved. Multiple remedial action projects may need to be completed before a site reaches construction completion (i.e., when all physical construction at a site is complete, all immediate threats have been addressed, and all long-term threats are under control). Removal actions are usually short-term cleanups for sites that pose immediate threats to human health or the environment. Examples of removal actions include removing and properly disposing of contaminated soil or other sources of hazardous materials (e.g., drum barrels) to prevent the release of hazardous substances, pollutants, or contaminants into the environment.

In prior reports,[5] we have provided information on the status of the nonfederal sites on the NPL. For example, based on our review of EPA data for our 2009 report, we determined that the number of nonfederal sites added to the NPL each year had on average declined from fiscal years 1983 to 2007. In addition, the types of nonfederal sites added to the NPL had also changed, as mining sites—among the most expensive sites to clean up—were added to the NPL in greater numbers during this period. We also determined that nonfederal NPL sites that had not yet reached construction completion may be more complex and costly to address. In our 2010 report, we found that federal appropriations to the Superfund program were generally declining, and limited funding had delayed remedial cleanup activities—which include remedial investigations, feasibility studies, and remedial action projects—at some nonfederal NPL sites.

You asked us to look at the current status of the cleanup of nonfederal NPL sites. This report examines, for fiscal years 1999 through 2013,[6] the trends in (1) the annual federal appropriations to the Superfund program and EPA expenditures on remedial cleanup activities at nonfederal sites on the NPL and (2) the number of nonfederal sites on the NPL, the number of remedial action project completions, and the number of construction completions at nonfederal NPL sites.

For the first objective, we reviewed and analyzed Superfund program funding and expenditure data for fiscal years 1999 through 2013. We obtained expenditure data from EPA's Integrated Financial Management System for fiscal years 1999 through 2003, and from its replacement financial system Compass, for fiscal years 2004 through 2013. For the second objective, we analyzed EPA data for nonfederal NPL sites for fiscal years 1999 through 2013. Specifically, we analyzed EPA data from the agency's Comprehensive Environmental Response, Compensation, and Liability Information System (CERCLIS) database to summarize trends in the number of new nonfederal sites added to the NPL, the number of nonfederal sites deleted from the NPL, the number of remedial action project completions, and the number of construction completions. The scope of our analyses for both objectives varied from year-to-year because we examined only nonfederal sites that were "active," i.e., on the NPL at any given point during the fiscal year. To address the objectives, we reviewed agency documents, including for example, the *Superfund Program Implementation Manual*, and interviewed EPA officials, including officials from Region 2, which includes the state that had the most nonfederal NPL sites in 2013. To assess the reliability of the data from EPA's databases used in this report, we reviewed relevant documents, such as the

2013 CERCLIS data entry control plan and regions' CERCLIS data entry control plans; examined the data to identify obvious errors or inconsistencies; compared the data that we received to publicly available data; and interviewed EPA officials. We determined the data to be sufficiently reliable for the purposes of this report. A more detailed discussion of our objectives, scope, and methodology is presented in appendix I.

We conducted this performance audit from October 2014 to September 2015 in accordance with generally accepted government auditing standards. Those standards require that we plan and perform the audit to obtain sufficient, appropriate evidence to provide a reasonable basis for our findings and conclusions based on our audit objectives. We believe that the evidence we obtained provides a reasonable basis for our findings and conclusions based on our audit objectives.

BACKGROUND

The Superfund process begins with the discovery of a potentially hazardous site or notification to EPA of the possible release of hazardous substances, pollutants, or contaminants that may threaten human health or the environment. EPA's regional offices may discover potentially hazardous waste sites, or such sites may come to EPA's attention through reports from state agencies or citizens. As part of the site assessment process, EPA regional offices use a screening system called the Hazard Ranking System to guide decision making, and as needed, to numerically assess the site's potential to pose a threat to human health or the environment. Those sites with sufficiently high scores are eligible to be proposed for listing on the NPL. EPA regions submit sites to EPA headquarters for possible listing on the NPL based on a variety of factors, including the availability of alternative state or federal programs that may be used to clean up the site. In addition, EPA officials have noted that, as a matter of policy, EPA seeks concurrence from the Governor of the state or state environmental agency head in which a site is located before listing the site. Sites that EPA proposes to list on the NPL are published in the Federal Register. After a period of public comment, EPA reviews the comments and decides whether to formally list the sites on the NPL.

EPA places sites into the following six broad categories based on the type of activity at the site that led to the release of hazardous material:

- Manufacturing sites include wood preservation and treatment, metal finishing and coating, electronic equipment, and other types of manufacturing facilities.
- Mining sites include mining operations for metals or other substances.
- "Multiple" sites include sites with operations that fall into more than one of EPA's categories.
- "Other" sites include sites that often have contaminated sediments or groundwater plumes with no identifiable source.
- Recycling sites include recycling operations for batteries, chemicals, and oil recovery.
- Waste management sites include landfills and other types of waste disposal facilities.

After a site is listed on the NPL, EPA or a potentially responsible party (PRP)[7] will generally begin the remedial cleanup process (see fig. 1) by conducting a two-part study of the site: (1) a remedial investigation to characterize site conditions and assess the risks to human health and the environment, among other actions, and (2) a feasibility study to evaluate various options to address the problems identified through the remedial investigation. The culmination of these studies is a record of decision (ROD) that identifies EPA's selected remedy for addressing the contamination. A ROD typically lays out the planned cleanup activities for each operable unit[8] of the site. EPA then plans the selected remedy during the remedial design phase, which is then followed by the remedial action phase when one or more remedial action projects are carried out. The number of operable units and planned remedial action projects at a site may increase or decrease over time as knowledge of site conditions changes. When all physical construction at a site is complete, all immediate threats have been addressed, and all long-term threats are under control, EPA generally considers the site to be construction complete. After construction completion, most sites then enter into the post-construction phase, which includes actions such as operation and maintenance during which the PRP or the state maintains the remedy such as groundwater restoration or a landfill cover, and EPA ensures that the remedy continues to protect human health and the environment. Eventually, when EPA and the state determine that no further site response is needed, EPA may delete the site from the NPL.

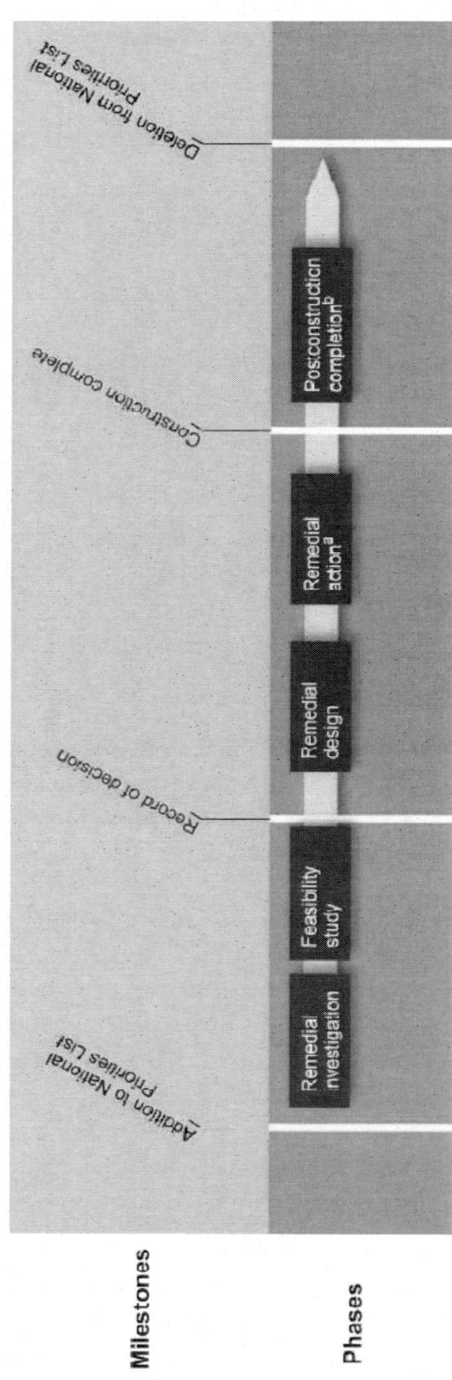

Source: GAO analysis of EPA data. I GAO-15-812.

Note: Phases of the remedial cleanup process may overlap, and multiple phases may be concurrently under way at a site.

[a]Remedial action projects occur during the remedial action phase; remedial action project completions also occur during this phase.

[b]Post-construction completion activities may include functions such as operation and maintenance, long-term response actions, and 5-year reviews, which ensure that Superfund cleanup actions provide for the long-term protection of human health and the environment.

Figure 1. Remedial Cleanup Process at National Priorities List Sites.

Measures to Communicate Physical Cleanup Progress

According to a 2000 *Federal Register* notice,[9] during the first 10 years of the Superfund program, the public often measured Superfund's progress in cleaning up sites by the number of sites deleted from the NPL as compared to the number of sites on the NPL. However, according to the same *Federal Register* notice, this measure did not recognize the substantial construction and reduction of risk to human health and the environment that had occurred at some NPL sites. In response, EPA established the sitewide construction completion measure to more clearly communicate to the public progress in cleaning up sites on the NPL. Similarly, according to EPA documents, in 2010, to augment the sitewide construction completion measure and reflect the amount of work being done at Superfund sites, EPA developed and implemented a new performance measure, remedial action project completions. EPA includes these two performance measures in its Annual Performance Plan.

General Funding Methods for Cleanup of Nonfederal NPL Sites

The cleanup of nonfederal NPL sites is generally funded by one or a combination of the following methods;

- Potentially responsible parties are liable for conducting or paying for site cleanup of hazardous substances.
- In some cases, PRPs cannot be identified or may be unwilling or financially unable to perform the cleanup. CERCLA authorizes EPA to pay for cleanups at sites on the NPL, including these sites. To fund EPA-led cleanups at nonfederal NPL sites, among other Superfund program activities, CERCLA established the Hazardous Substance Superfund Trust Fund (Trust Fund). Historically, the Trust Fund was financed primarily by taxes on crude oil and certain chemicals, as well as an environmental tax on corporations. The authority to levy these taxes expired in 1995. Since fiscal year 2001, appropriations from the general fund have constituted the largest source of revenue for the Trust Fund. About 80 percent of the funds EPA spent to clean up nonfederal NPL sites from 1999 through 2013 came from annual appropriations. The remaining roughly 20 percent came from special

accounts and state cost share. EPA has limited cost data where a PRP has conducted the cleanup.[10]

- Under CERCLA, EPA is authorized to enter into settlement agreements with PRPs to pay for cleanups, and EPA may retain and use these funds for cleanups. Funds from these settlements may be deposited into site-specific subaccounts in the Trust Fund, which are referred to as "special accounts" and are generally used for future cleanup actions at the sites associated with a specific settlement, or to reimburse funds that EPA had previously used for response activities at these sites. According to EPA documents, in fiscal year 2013, there were a total of 993 open special accounts with an end of year balance of about $1.7 billion. Most of these funds could be used for a limited number of sites—for example, 3 percent of the open accounts representing 33 sites had about 56 percent of the total special account resources available.

- States are required to pay 10 percent of Trust Fund-financed remedial action cleanup costs and at least 50 percent of cleanup costs for facilities that were operated by the state or any political subdivision of the state at the time of any hazardous substances disposal at the facility. States may pay their share of response costs using cash, services, credit, or any combination thereof. Under CERCLA, states are also required to assure provision of all future maintenance of a Trust Fund-financed remedial action.

EPA's New Information Management System

In fiscal year 2014, EPA updated its information system for the Superfund program from CERCLIS to the Superfund Enterprise Management System (SEMS). According to EPA officials and documents, SEMS consolidated five stand-alone information systems and reporting tools into one system. These systems include CERCLIS, the Superfund Document Management System (SDMS), the Institutional Controls Tracking System (ICTS), the eFacts reporting tool, and ReportLink. CERCLIS contained information on, among other things, the contaminated sites' cleanup status and cleanup milestones reached. The SDMS was a national electronic records collection system mostly with site cleanup records;

ICTS was a database with legal data related to controlling access to sites; eFacts was a visual reporting tool that generated charts and graphs; and

ReportLink was a traditional reporting tool that allowed regions and headquarters to share reports. According to EPA officials, SEMS should be more user-friendly and provide more mobility, thus allowing EPA regional staff to access the system in the field through various devices. Currently, regions are in the process of entering data for each site into SEMS. The process of converting entirely to SEMS has taken additional time because, according to EPA officials, the complexity of the new software and its difference from CERCLIS has created a more significant obstacle than anticipated. In addition, EPA officials stated that the agency will not be in a position to release data comparable to the data previously shared from CERCLIS until EPA officials are confident that all regions have mastered the software to update site schedules. According to EPA officials, SEMS should be fully operable in fiscal year 2016.

Estimated U.S. Population Living within 3 Miles of a Nonfederal NPL Site

According to our analysis of EPA and Census data,[11] as of fiscal year 2013, an estimated 39 million people—about 13 percent of the U.S. population—lived within 3 miles of a nonfederal NPL site. Many of these people—an estimated 14 million—were either under the age of 18 or 65 years and older, which EPA describes as sensitive subpopulations. EPA Region 2 had the largest number of people living within 3 miles of a nonfederal NPL site—an estimated 10 million or about one-third of the region's total population. Figure 2 provides information on the number of nonfederal NPL sites in each region and the estimated number of people that lived within 3 miles of those sites as of fiscal year 2013. The state of New York had the largest number of people living within 3 miles of nonfederal NPL sites—an estimated 6 million or about 29 percent of the state's population. The state of New Jersey had the largest percentage of its estimated population living within 3 miles of a nonfederal NPL site— about 50 percent. Appendix II provides information on the estimated population that lived within 3 miles of a nonfederal NPL site, by state, as of fiscal year 2013.

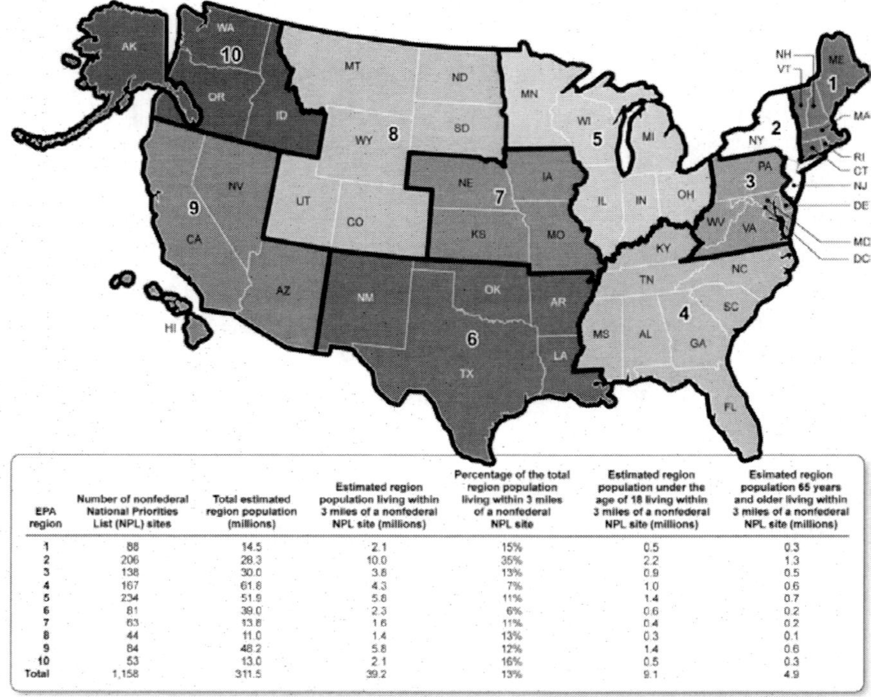

EPA region	Number of nonfederal National Priorities List (NPL) sites	Total estimated region population (millions)	Estimated region population living within 3 miles of a nonfederal NPL site (millions)	Percentage of the total region population living within 3 miles of a nonfederal NPL site	Estimated region population under the age of 18 living within 3 miles of a nonfederal NPL site (millions)	Esimated region population 65 years and older living within 3 miles of a nonfederal NPL site (millions)
1	88	14.5	2.1	15%	0.5	0.3
2	206	28.3	10.0	35%	2.2	1.3
3	138	30.0	3.8	13%	0.9	0.5
4	167	61.8	4.3	7%	1.0	0.6
5	234	51.9	5.8	11%	1.4	0.7
6	81	39.0	2.3	6%	0.6	0.2
7	63	13.8	1.6	11%	0.4	0.2
8	44	11.0	1.4	13%	0.3	0.1
9	84	48.2	5.8	12%	1.4	0.6
10	53	13.0	2.1	16%	0.5	0.3
Total	1,158	311.5	39.2	13%	9.1	4.9

Source: GAO analysis of EPA data and U.S. Census data; Map Resources (map). | GAO-15-812.

Note: The methodology for GAO's analysis is generally based on EPA's, Office of Solid Waste and Emergency Response's approach. Data analyzed include (1) 1,158 nonfederal sites on the National Priorities List, in the 50 states and U.S. territories (Guam, Puerto Rico, and the Virgin Islands), as of the end of fiscal year 2013, and (2) Census data from the 2009-2013 American Community Survey 5-year estimate for the 1,141 nonfederal sites in the 50 states and the District of Columbia. A circular site boundary, equal to the site acreage, was modeled around the latitude/longitude for each site and then a 3-mile buffer ring was placed around the site boundary. For the 138 sites in 34 states that EPA did not have acreage information, a circular site boundary was modeled around the latitude/longitude point, and then a 3-mile buffer ring was placed around the point. American Community Survey data was then collected for each block group with a centroid that fell within the 3-mile area and rounded. Percentage numbers were rounded to the nearest whole percent.

Figure 2. Estimated U.S. Population Living within 3 Miles of Nonfederal National Priorities List Sites by EPA Region, as of Fiscal Year 2013.

ANNUAL FEDERAL SUPERFUND APPROPRIATIONS DECREASED, AND EPA EXPENDITURES ON REMEDIAL CLEANUP ACTIVITIES DECLINED

Annual federal appropriations (appropriations) to EPA's Superfund program generally declined from about $2 billion to about $1.1 billion from fiscal years 1999 through 2013. EPA expenditures—from these federal appropriations—of site-specific cleanup funds (funds spent on remedial cleanup activities at nonfederal NPL sites) declined from about $0.7 billion to about $0.4 billion during the same time period. Because EPA prioritizes funding work that is ongoing, the decline in funding led EPA to delay the start of about one-third of the new remedial action projects that were ready to begin in a given fiscal year at nonfederal NPL sites from fiscal years 1999 through 2013, according to EPA officials. EPA spent the largest amount of cleanup funds in Region 2, which accounted for about 32 percent of cleanup funds spent at nonfederal NPL sites from fiscal years 1999 through 2013. During the same time period, EPA spent the majority of cleanup funds in seven states, with the most in New Jersey— over $2.0 billion or more than 25 percent of cleanup funds. According to our analysis of EPA data, the median per-site annual expenditures for cleanup at nonfederal NPL sites declined by about 48 percent from fiscal years 1999 through 2013, and EPA spent the majority of cleanup funds on an average of about 18 sites annually. Unless otherwise indicated, all dollars and percentage calculations are in constant 2013 dollars.

Annual Federal Appropriations for the Superfund Program and EPA Expenditures of Site-Specific Cleanup Funds on Remedial Cleanup Activities Decreased

From fiscal years 1999 through 2013, the annual appropriations to EPA's Superfund program generally declined. Annual appropriations declined from about $2 billion to about $1.1 billion—about 45 percent—from fiscal years 1999 through 2013.[12] Under the American Recovery and Reinvestment Act of 2009 (Recovery Act),[13] EPA's Superfund program received an additional $639 million in fiscal year 2009.[14] Figure 3 shows the annual federal appropriations from fiscal years 1999 through 2013.

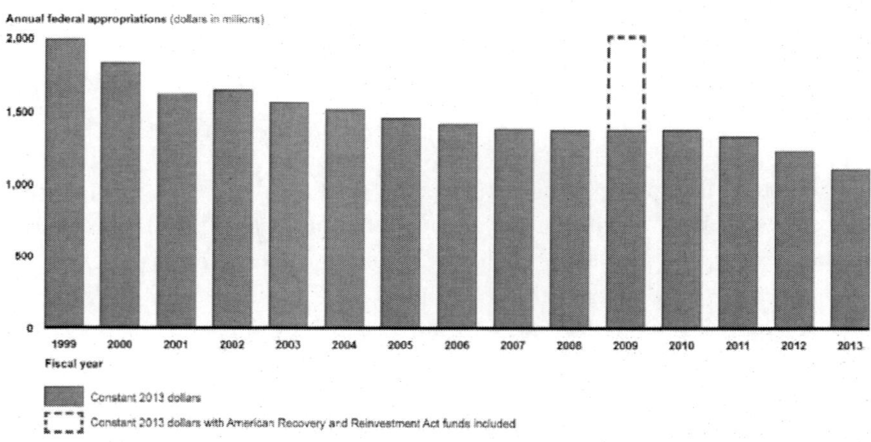

Source: GAO analysis of data from the President's Budget Appendixes. | GAO-15-812.

Figure 3. EPA's Superfund Program Annual Appropriations, Fiscal Years 1999 through 2013.

EPA allocates annual appropriations to the Superfund program among the remedial program and other Superfund program areas, such as enforcement (see fig. 4). The remedial program generally funds cleanups of contaminated nonfederal NPL sites. EPA headquarters allocates funds for the remedial program to various categories: payroll and other administrative activities; preconstruction and other activities (such as remedial investigations and feasibility studies); and construction (such as remedial action projects) and post-construction activities. EPA allocates funds for preconstruction and other activities to its regional offices using a model based on a combination of historical allocations and a scoring system based on regions' projects planned for the upcoming year. Each region decides how it will spend funds allocated by headquarters for its preconstruction and other remedial activities. EPA headquarters, in consultation with the regions, allocates site-specific cleanup funds for construction and post-construction activities between ongoing work and new remedial action projects.

From fiscal years 1999 through 2013, the decline in appropriations to the Superfund program led EPA to decrease expenditures of site-specific cleanup funds on remedial cleanup activities from about \$0.7 billion to about \$0.4 billion.[15] We define site-specific cleanup funds as those funds spent on preconstruction, construction, and postconstruction, which comprise remedial cleanup activities. Expenditures of Recovery Act funds account for the increase in cleanup funds expenditures from fiscal years 2009 through 2011.[16]

Figure 5 shows EPA's expenditures of cleanup funds at nonfederal NPL sites for fiscal years 1999 through 2013.

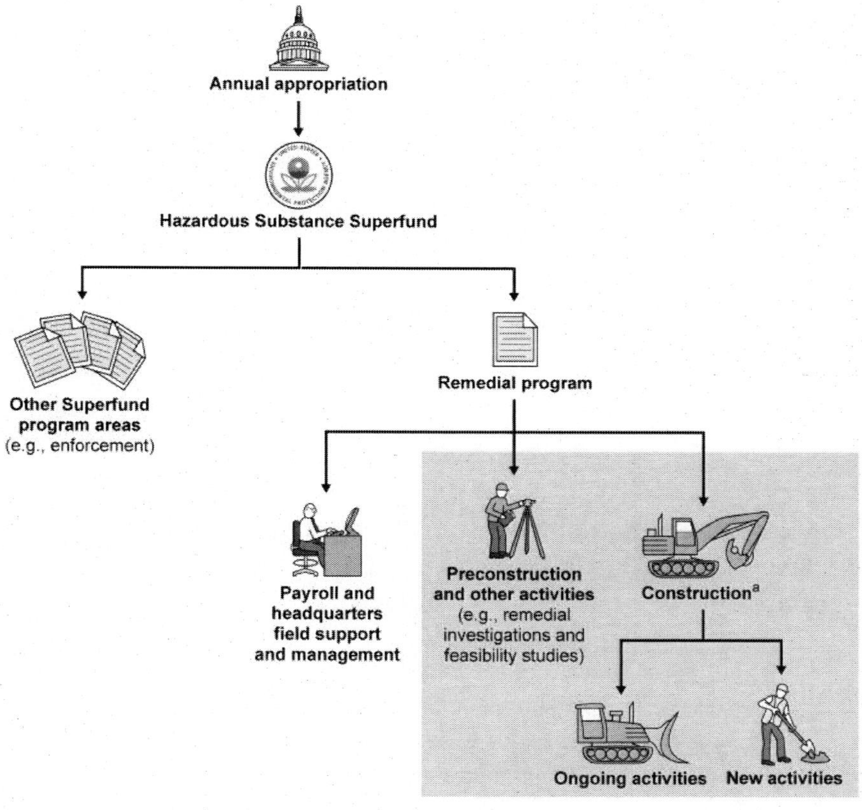

Source: GAO analysis of EPA information. | GAO-15-812.

Construction includes post-construction activities, such as long-term response actions and 5-year reviews.

Figure 4. EPA's Process for Allocating Annual Federal Appropriations to the Superfund Program.

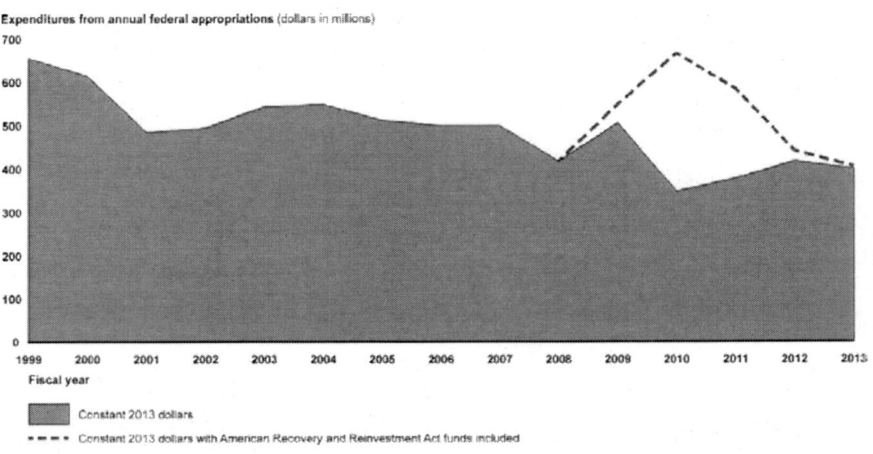

Source: GAO analysis of EPA data. | GAO-15-812.

Figure 5. EPA Expenditures of Site-Specific Cleanup Funds on Remedial Cleanup Activities at Nonfederal National Priorities List Sites, Fiscal Years 1999 through 2013.

EPA's Prioritization of Ongoing Work in the Context of Decreased Funding Delayed the Start of Some New Remedial Action Projects

EPA policy prioritizes funding ongoing work over starting new remedial action projects. EPA officials explained that funding ongoing work is prioritized for a variety of reasons, such as the risk of recontamination and the additional cost of demobilizing and remobilizing equipment and infrastructure at a site. To establish funding priorities for new remedial action projects, EPA's National Risk-Based Priority Panel (Panel)— comprised of EPA regional and headquarters program experts—ranks new remedial action projects based on their relative risk to human health and the environment. The Panel uses five criteria to evaluate proposed new remedial action projects: (1) risks to human population exposed (e.g., population size and proximity to contaminants), (2) contaminant stability (e.g., use and effectiveness of institutional controls like warning signs), (3) contaminant characteristics (e.g. concentration and toxicity), (4) threat to a significant environmental concern (e.g., endangered species or their critical habitat), and (5) program management considerations (e.g., high-profile projects). Each criterion is ranked on a weighted scale of one to five with the highest score for any criterion being five. According to EPA documents, the priority ranking process ensures that funding decisions for new remedial action projects are based on common evaluation criteria that emphasize risk to human health and the environment. The Panel then recommends the new projects to fund to the Assistant Administrator of the Office of Solid Waste and Emergency Response who makes the final funding decisions.[17]

A decline in funding delayed the start of some new remedial action projects, according to EPA officials. Over the 15-year time period from fiscal years 1999 through 2013, EPA generally did not fund all of the new remedial action projects that were ready to begin in a given fiscal year, according to our analysis of EPA data, (see table 1). During this time, EPA did not fund about one-third of the new remedial action projects in the year in which they were ready to start. According to EPA officials in headquarters and Region 2, delays in starting new remedial action projects can potentially lead to elevated costs. For example, site conditions can change, such as contaminants migrating at a groundwater site, which will require recharacterization of the location. Also the extent of the contamination may change or adjustments may be necessary to the remedy designs which could take additional time and money. In addition, there may be unmeasured economic costs to the community by delaying the productive reuse of a site, according to EPA officials. Due to an increase in funding from the Recovery Act, EPA started all new remedial action projects ready to start in fiscal years 2009 and 2010, and most new remedial action projects in fiscal year 2011, according to our analysis of EPA data. However, in fiscal year 2012, EPA did not fund and start any of the 21 new remedial action projects through the Panel process[18] that were ready to begin that year. The 21 unfunded projects were estimated to have cost over $117 million[19] in 2012, according to EPA officials. In fiscal year 2013, EPA did not fund 22 out of 30 projects due to priorities for declining funds as well. According to EPA officials, in that year, these unfunded projects were estimated to have cost approximately $101 million. EPA officials stated that they expect the trend of being unable to fund all new remedial action projects to continue.[20] According to EPA officials, prior to funding new remedial action projects, EPA considers both the funds needed in the current fiscal year to begin the project and ongoing funds that will be required in subsequent fiscal years to complete the project.

Table 1. Annual Funding Decisions for New Remedial Action Projects at EPA's Nonfederal National Priorities List Sites Ranked by the National Risk-Based Priority Panel, Fiscal Years 1999 through 2013

Fiscal year	Projects funded	Projects not funded	Total projects	Percentage projects funded	Percentage projects not funded
1999	16	0[a]	16	100%	0%
2000	15	12	27	56%	44%
2001	4	16	20	20%	80%
2002	17	7	24	71%	29%
2003	9	12	21	43%	57%
2004	27	19	46	59%	41%
2005	17	9	26	65%	35%
2006	18	6	24	75%	25%
2007	19	0	19	100%	0%
2008	16	10	26	62%	38%
2009	26	0	26	100%	0%
2010	18	0	18	100%	0%
2011	12[b]	4	16	75%	25%
2012	0[b]	21	21	0%	100%
2013	8[b]	22	30	27%	73%

Source: GAO analysis of EPA data. | GAO-15-812

Note: EPA's National Risk-Based Priority Panel (Panel) does not rank new remedial action projects financed entirely by special account funds. The numbers in each fiscal year represent funding decisions for new remedial action projects ready to begin construction during that particular fiscal year. According to EPA officials, in total, funding for 94 discrete new remedial action projects was delayed—34 percent of the total number of remedial action projects considered for funding during fiscal years 1999 through 2013. Percentage numbers were rounded to the nearest whole percent.

[a]We did not include 20 remedial action projects that were presented to the Panel in fiscal year 1999 because, according to EPA officials, the agency could not differentiate between those projects that were not funded, and those removed from consideration because they were not construction-ready in that fiscal year.

[b]The number of new remedial action projects funded in fiscal years 2011, 2012, and 2013 does not match the number EPA reports publically in its annual accomplishment reports due to reporting differences. For EPA accomplishment reports, see http://www.epa.gov/superfund/accomplishments.htm (accessed July 23, 2015).

According to EPA officials, as annual appropriations have declined, EPA has generally relied on funds available from prior year Superfund appropriations to fund new remedial action projects and some other work. According to EPA officials, funds from prior year appropriations generally become available for use through deobligations and special account reclassifications. Typically, deobligations occur when EPA determines that some or all of the funds the agency originally obligated for a contract to conduct an activity are no longer needed (e.g., EPA will deobligate funds that it had previously obligated to construct a landfill cover because the final costs were less than originally anticipated). According to EPA officials, reclassifications occur when EPA uses special account funds to reimburse itself for its past expenditures of annually appropriated funds, which then makes the funds originally used for these activities available for the agency to use. Starting in fiscal year 2003, EPA began distributing deobligated funds in a 75/25 percent split so that headquarters kept 75 percent of the deobligated funds for national remedial program priorities, which have been, in large part, used to begin new remedial action projects, and returned 25 percent to the region that provided the deobligated funds. On average, EPA annually provided about $58 million in deobligated funds for construction and post-construction activities during fiscal years 2003 through 2013,[21] according to our analysis of EPA data. According to EPA officials, deobligations are an unpredictable funding stream, and our analysis of EPA data indicates that the amount of deobligations and reclassifications provided for cleanup fluctuated during the fiscal years 2003 through 2013 time period, from a high in fiscal year 2003 of about $102 million to a low in fiscal year 2009 of about $32 million.

EPA Spent the Most Cleanup Funds in One Region, and Primarily in Seven States

EPA spent the most cleanup funds from annual appropriations on nonfederal NPL sites in Region 2 from fiscal years 1999 through 2013, according to our analysis of EPA data. EPA spent almost $2.5 billion in this region—which is about 32 percent of the total cleanup funds on nonfederal NPL sites during that time frame and over three times the cleanup funds spent on any other region.[22] According to EPA officials, Region 2 has a significant number of large, EPA-funded sites that have required considerable expenditures to clean up over a long period of time. The agency does not expect this trend to continue, but anticipates that more cleanup funds will be devoted to the cleanup of large mining and sediment sites in the West. Region 8 received the second most in cleanup funds with about $0.7 billion over the same time period. Figure 6 shows EPA's expenditure of cleanup funds at nonfederal NPL sites in each region from fiscal years 1999 through 2013.

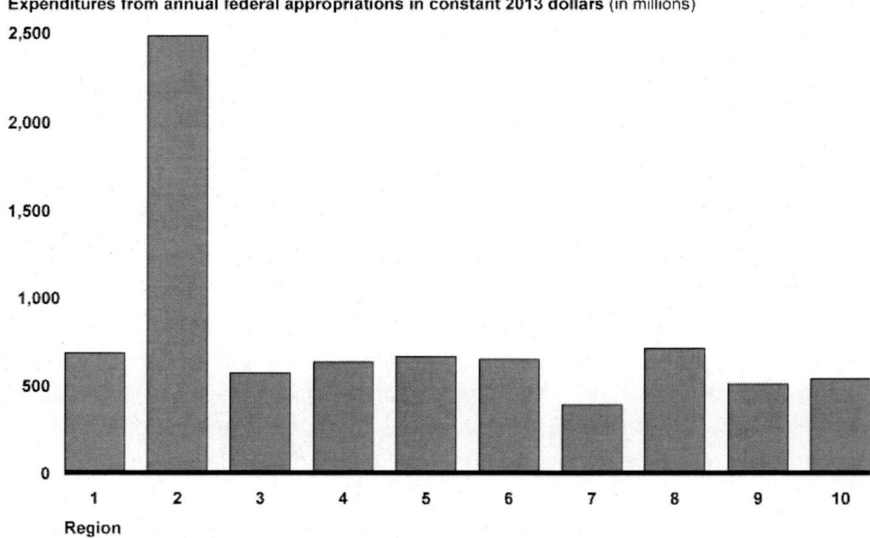

Source: GAO analysis of EPA data. | GAO-15-812

Figure 6. EPA Expenditures of Site-Specific Cleanup Funds on Remedial Cleanup Activities at Nonfederal National Priorities List Sites, by Region, Fiscal Years 1999 through 2013.

According to our analysis of EPA data, EPA spent the majority of nonfederal NPL cleanup funds in seven states—New Jersey, California, New York, Massachusetts, Idaho, Pennsylvania, and Florida—during the 15-year period from fiscal years 1999 through 2013. New Jersey sites received the most cleanup funds with over $2.0 billion (or more than 25 percent of cleanup funds over this period).[23] The agency also spent the largest portion of Recovery Act funds in New Jersey. According to EPA officials, New Jersey has a large number of sites that do not have PRPs to perform the cleanup and needs federal appropriations to cleanup these sites.[24] In addition, sites in areas of highly dense population like many in New Jersey cost more to cleanup, according to EPA officials. Agency officials expect the current level of expenditures in New Jersey to decline in the future because the cleanup at some of the sites will be completed. Figure 7 shows EPA's expenditure of cleanup funds in the seven states from fiscal years 1999 through 2013.

Expenditures from annual federal appropriations in constant 2013 dollars (in millions)

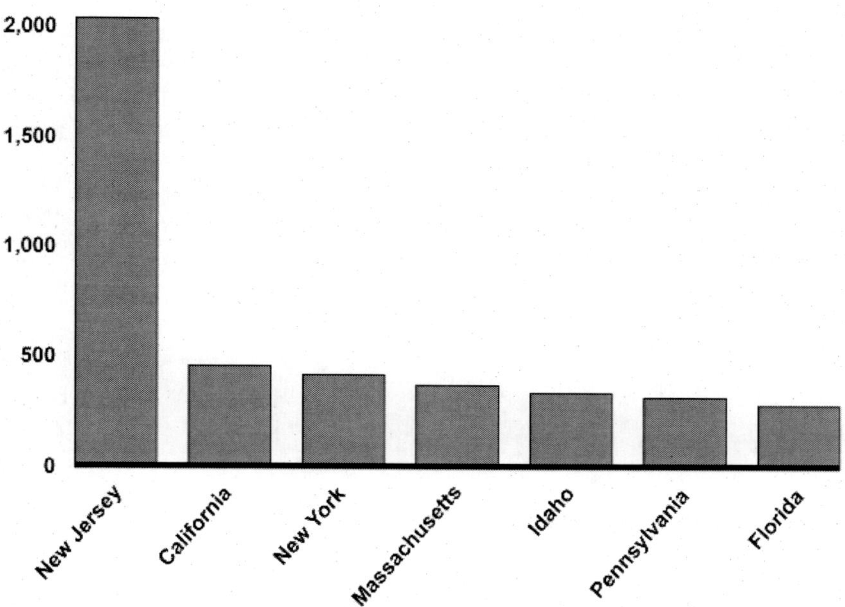

Source: GAO analysis of EPA data. | GAO-15-812

Figure 7. EPA Expenditures of Site-Specific Cleanup Funds on Remedial Cleanup Activities at Nonfederal National Priorities List Sites in the States Where EPA Spent the Most Cleanup Funds, Fiscal Years 1999 through 2013.

Median Annual Expenditures per Site Declined and on Average, EPA Spent the Majority of Cleanup Funds at 18 Sites Annually

According to our analysis of EPA data, the median per-site annual expenditures on remedial cleanup activities at nonfederal NPL sites generally declined from fiscal years 1999 through 2013.[25] The median per-site annual expenditures declined by about 48 percent from about $36,600 to about $19,100 from fiscal years 1999 through 2013.[26] The decline was more pronounced in recent years, decreasing by about 35 percent from fiscal years 2009 through 2013, compared to about a 12 percent decline from fiscal years 1999 through 2003. Figure 8 shows the median per-site annual expenditures of cleanup funds from annual appropriations at nonfederal NPL sites from fiscal years 1999 through 2013. According to EPA officials, these declines mirror, with some lag time, declines in appropriations, the most significant of which occurred starting in fiscal year 2000 and then again starting in fiscal year 2011. In addition, the agency expects to see further declines in annual cleanup funds expenditures following the same pattern in the near future, according to EPA officials. Specifically, given recent declines in appropriations, EPA expects to see declines in expenditures after a short lag time, while outyear[27] trends would depend on future appropriations.

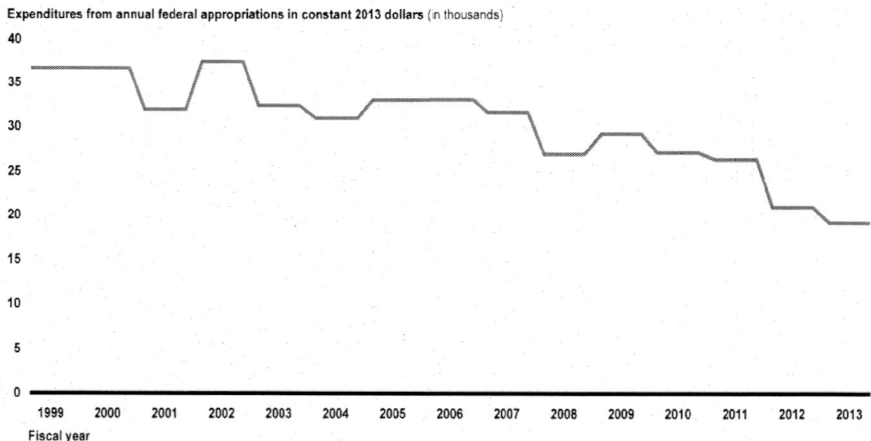

Source: GAO analysis of EPA data. | GAO-15-812

Figure 8. EPA's Median Per-Site Annual Expenditures of Site-Specific Cleanup Funds on Remedial Cleanup Activities at Nonfederal National Priorities List Sites, Fiscal Years 1999 through 2013.

EPA spent the majority of cleanup funds on a few sites—on average about 18 sites—each year from fiscal years 1999 through 2013, according to our analysis of EPA data. The specific sites where EPA spent the majority of cleanup funds varied from year to year, but 6 sites were part of the 18 in more than half the years of the 15-year period— Vineland Chemical Company, Inc. (New Jersey), Bunker Hill Mining and Metallurgical Complex (Idaho), Welsbach and General Gas Mantle-Camden Radiation (New Jersey), Tar Creek-Ottawa County (Oklahoma), New Bedford (Massachusetts), and

Federal Creosote (New Jersey). EPA spent at least $175 million from annual appropriations at each of these 6 sites over the 15 years.

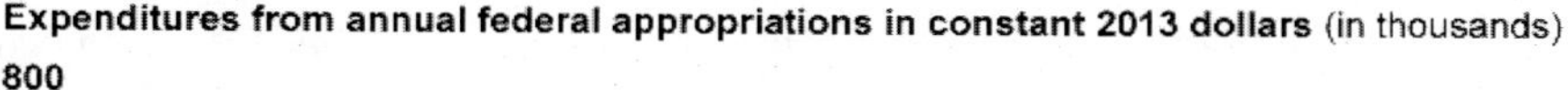

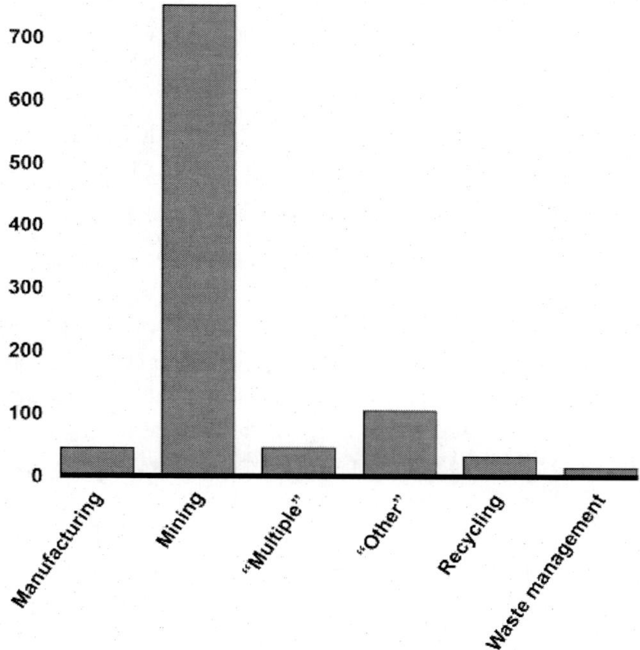

Source: GAO analysis of EPA data. | GAO-15-812

Figure 9. EPA's Average Median Per-Site Annual Expenditure of Site-Specific Cleanup Funds on Remedial Cleanup Activities at Nonfederal National Priorities List Sites, by Site Type, Fiscal Years 1999 through 2013.

EPA's costs to clean up sites differed depending on the type of site. According to our analysis of EPA data on expenditures of cleanup funds from annual appropriations, mining sites were the most expensive to clean up. From fiscal year 1999 through 2013, EPA spent, on average, from about 7 to about 52 times the annual amount per site at mining sites than at the other types of sites. For example, the average median per-site annual expenditure of cleanup funds was about $750,000 for mining sites compared to about $104,000 for "other" sites and to about $14,000 for waste management sites. According to EPA officials, mining sites are costly to clean up because, among other characteristics, they typically cover a large area and have many sources of contamination. One example of a mining site is the Bunker Hill Mining and Metallurgical Complex in Idaho where EPA spent almost $330 million to clean up part of the site from fiscal years 1999 through 2013.[28] Figure 9 shows the average median per-site annual expenditure of cleanup funds from annual appropriations at nonfederal NPL sites by type of site from fiscal years 1999 through 2013.

THE NUMBER OF NONFEDERAL SITES ON THE NPL REMAINED RELATIVELY CONSTANT, WHILE THE NUMBER OF REMEDIAL ACTION PROJECT COMPLETIONS AND CONSTRUCTION COMPLETIONS GENERALLY DECLINED

According to our analysis of EPA data, the total number of nonfederal sites on the NPL annually remained relatively constant, while remedial action project completions and construction completions generally declined during fiscal years 1999 through 2013.[29] The total number of nonfederal sites on the NPL increased from 1,054 in fiscal year 1999 to 1,158 in fiscal year 2013 and averaged about 1,100 annually. According to our analysis of EPA data, the number of remedial action project completions at nonfederal NPL sites generally declined by about 37 percent during the 15-year period. Similarly, from fiscal years 1999 through 2013, the number of construction completions at nonfederal NPL sites generally declined by about 84 percent.

The Total Number of Nonfederal Sites on the NPL Remained Relatively Constant

From fiscal years 1999 through 2013, the number of new nonfederal sites added to the NPL and the number of nonfederal sites deleted each year from the NPL generally declined, while the total number of nonfederal sites on the NPL remained relatively constant, according to our analysis of EPA data. More specifically, during the fiscal years of our review, there was a period of decline in the number of sites added to the NPL followed by a few years where there was a slight increase. For example, the number of new nonfederal sites added to the NPL each year declined steadily from 37 sites in fiscal year 1999 to 12 in fiscal year 2007. According to EPA officials, there are several reasons for the decline in the number of new nonfederal sites added to the NPL. For example, some states may have been managing the cleanup of sites with their own state programs, especially if a PRP was identified to pay for the cleanup.[30] Additional reasons for the decrease during this time period include:

(1) funding constraints that led EPA to focus primarily on sites with actual human health threats and no other cleanup options, (2) use of the NPL as a mechanism of last resort, and (3) referral of sites assessed under Superfund to state cleanup programs.

In contrast, from fiscal years 2008 through 2012, there was a general increase in the number of new nonfederal sites added to the NPL annually, according to our analysis of EPA data. In fiscal year 2008, EPA added 18 sites and by 2012, the number of sites added annually had increased to 24. According to EPA officials, the numbers may have increased from fiscal years 2008 through 2012, because the agency expanded its focus to consider NPL listing for sites with potential human health and environmental threats, and it shifted its policy to use the NPL when it was deemed the best approach for achieving site cleanup rather than using the NPL as a mechanism of last resort. Also, states' funding for cleanup programs declined, and states agreed to add sites to the NPL where they encountered difficulty in getting a PRP to cooperate or where the PRP went bankrupt, according to EPA officials. Furthermore, these same officials stated that the increase in the number of new sites added to the NPL could be due to referrals from the Resource Conservation and Recovery Act program because of business bankruptcies, especially in the most recent years.[31] In fiscal year 2013, however, the number of new nonfederal sites added to the NPL declined to 8, the lowest number since fiscal year 1999. In total, EPA added 304 nonfederal sites to the NPL— an average of about 20 sites annually—from fiscal years 1999 through 2013.[32] Figure 10 summarizes the number of new nonfederal sites added to the NPL each year from fiscal years 1999 through 2013.

In terms of the types of sites added to the NPL from fiscal years 1999 through 2013, the largest number of sites added to the list were manufacturing sites (120 sites or about 40 percent) followed by "other" sites (90 sites or about 30 percent). In addition, EPA added 35 mining sites (about 12 percent), 32 waste management sites (about 11 percent), 21 recycling sites (about 7 percent), and 6 "multiple" sites (about 2 percent)—sites that fell into more than one of these categories— according to our analysis of EPA data. During this time frame, the amount of time between when a site was proposed to be added to the NPL and when it was added to the NPL ranged from 2 months to over 18 years, with a median amount of time of about 6 months.[33] According to EPA officials, there are a variety of reasons to explain why some sites take longer to add to the NPL. For example, EPA could propose a site to be added to the NPL and, in response to the *Federal Register* notice announcing the proposal, EPA could receive numerous, complex comments that required considerable time and EPA resources to address. In addition, a proposal to add a site to the NPL could act as an incentive for PRPs to resume negotiations with EPA or the state to clean up the site.[34] Moreover, large PRPs with greater financial assets may request additional time to pursue other cleanup options; hire law firms and technical contractors to submit challenging comments to EPA on the proposal to add the site to the NPL; and support outreach efforts that generate state and local opposition to the proposal. EPA officials also noted that certain sites, such as recycling and dry cleaning,[35] are generally added quickly to the NPL because other alternatives may not be available.

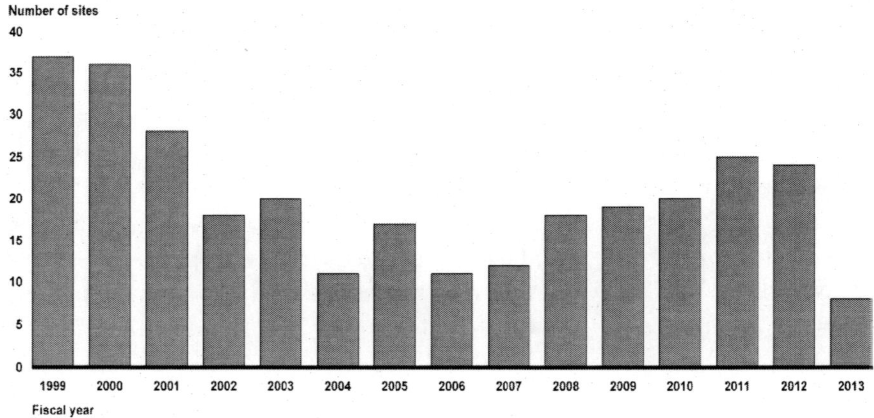

Source: GAO analysis of EPA data. | GAO-15-812

Figure 10. Number of New Nonfederal Sites Added to the National Priorities List Each Year, Fiscal Years 1999 through 2013.

From fiscal years 1999 through 2013, the number of nonfederal sites deleted from the NPL generally declined, according to our analysis of EPA data. EPA deleted 22 nonfederal sites in fiscal year 1999 and, in fiscal year 2013, EPA deleted only 6 nonfederal sites. In total, EPA deleted 185 nonfederal sites from the NPL during these years.[36] According to EPA officials, the decline in the number of nonfederal sites deleted from the NPL is due to the decline in annual appropriations and the fact that the sites remaining on the NPL are more complex, and they take more time and money to clean up. The median number of years from the time a nonfederal site was added to the NPL to the time EPA deleted it from the NPL ranged from about 13 years for those sites deleted in fiscal year 1999, to about 25 years for those sites deleted in fiscal year 2013, with an average median of about 19 years. Region 2 had the largest number of nonfederal sites—41—deleted from the NPL, followed by Regions 6, 3, 4, and 5, which deleted 29, 25, 23, and 23 nonfederal sites, respectively. Figure 11 shows the number of nonfederal sites EPA deleted from the NPL each year from fiscal years 1999 through 2013.

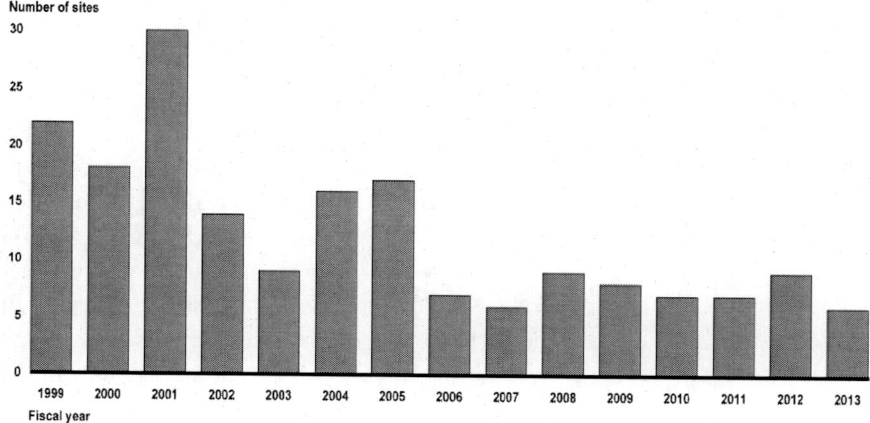

Source: GAO analysis of EPA data. | GAO-15-812

Figure 11. Number of Nonfederal Sites Deleted from the National Priorities List Each Year, Fiscal Years 1999 through 2013.

From fiscal years 1999 through 2013, according to our analysis of EPA data, the total number of nonfederal sites on the NPL remained relatively constant, and averaged about 1,100 sites annually. From fiscal years 1999 through 2013, the total number of nonfederal sites on the NPL increased less than 10 percent—from 1,054 sites to 1,158 sites as of the end of these fiscal years.[37] In addition, the type of nonfederal sites on the NPL changed during this same time period. For example, in fiscal year 1999, there were 10 mining sites on the NPL or about 1 percent of all nonfederal NPL sites. By fiscal year 2013, there were 44 mining sites on the NPL, which was about 4 percent of all nonfederal NPL sites. Appendix III provides more detailed information from fiscal years 1999 through 2013 on the number of nonfederal sites on the NPL at the end of each fiscal year, following any additions and deletions; as well as the number of nonfederal sites on the NPL each fiscal year by type.

Remedial Action Project Completions and Construction Completions Generally Declined

According to our analysis of EPA data, from fiscal years 1999 through 2013, the number of remedial action project completions at nonfederal NPL sites declined by about 37 percent, and the length of time to complete the projects increased slightly. The number of remedial action project completions in each year gradually declined by about 59 percent from 116 projects (fiscal year 1999) to 47 projects (fiscal year 2010). For fiscal years 2011 through 2012, the number of remedial action project completions increased to 75 and 87, respectively. According to EPA officials, these increases were due to the increase of funds from the Recovery Act. In fiscal year 2013, the number of remedial action project completions dropped to 73. In total, 1,181 remedial action projects were completed from fiscal years 1999 through 2013.[38] In general, according to EPA officials, the decline in remedial action project completions is due to the decline in appropriations and the complexity of current projects, which take longer to complete. These officials also stated that the decline in staffing, especially in the last few years, and particularly in the regions, had a negative impact on the Superfund remedial program and made it difficult to complete work. Figure 12 provides information on the number of remedial action project completions at nonfederal NPL sites from fiscal years 1999 through 2013.

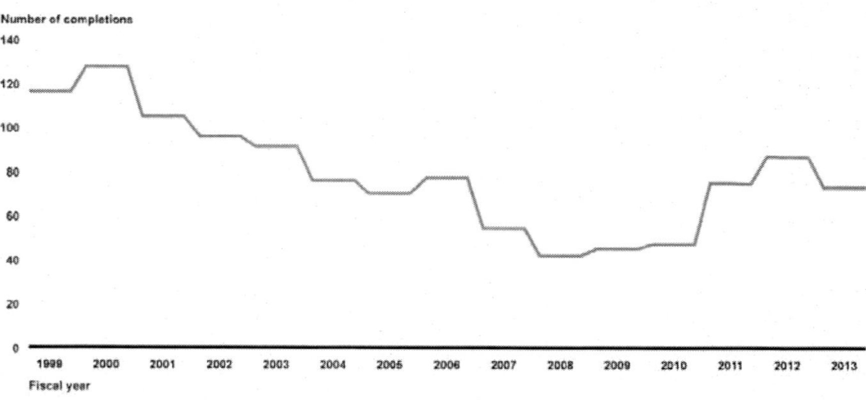

Source: GAO analysis of EPA data. | GAO-15-812

Figure 12. Number of Remedial Action Project Completions at Nonfederal National Priorities List Sites, Fiscal Years 1999 through 2013.

According to our analysis of EPA data, Region 2 had the highest number of remedial action project completions (242 projects or about 20 percent of the total project completions), followed by Regions 3, 5, and 4 at 171 projects (or about 14 percent), 140 projects (or about 12 percent), and 128 projects (or about 11 percent), respectively. New Jersey, Pennsylvania, and New York completed the most remedial action projects—over 100 projects in each state—during the 15-year time frame.

In addition to fewer remedial action project completions, our analysis of EPA data also shows that the length of time to complete these projects increased slightly from one year to the next. From fiscal years 1999 through 2013, the average median length of time to complete these projects was about 3 years. In fiscal year 1999, the median amount of time to complete projects was about 2.6 years. Over time, the median amount of time gradually increased to almost 4 years in fiscal year 2013. Regions 6 and 3 had the lowest average median times of about 2 years to complete projects. In contrast, Region 10 had the highest average median time of over 5 years to complete projects. According to EPA officials, remedial action project completions are taking longer to complete because they are getting more complex. In addition, these officials stated that, as noted above, shortages in EPA regional staffing levels and a decline in state environmental agency personnel are causing delays throughout the Superfund program from site assessments to completion of remedial action projects.

Similar to the decline in the number of remedial action project completions, from fiscal years 1999 through 2013, the number of construction completions at nonfederal NPL sites generally declined by about 84 percent, according to our analysis of EPA data. Specifically, fiscal years 1999 and 2000 had the largest number of construction completions at nonfederal NPL sites— 80 sites each fiscal year. In contrast, in fiscal year 2013, the number of construction completions at nonfederal NPL sites declined to 13. During the 15-year time frame, 516 nonfederal NPL sites reached construction completion.[39] According to EPA officials, the decline in the number of construction completions at nonfederal NPL sites is because, as noted above, the sites are getting more complex and difficult to clean up, funds available to perform the cleanup are declining, the number of sites available for construction completion have declined from fiscal years 1999 through 2013, and regional staff is declining. In addition, adverse weather conditions, such as excessive rain, and the discovery of new contaminants can delay progress at some sites, according to these same officials. Figure 13 shows the trend in the number of construction completions at nonfederal NPL sites from fiscal years

1999 through 2013. In fiscal year 1999, the median number of years to reach construction completion was about 12 years, and in fiscal year 2013, it was about 16 years. During the 15-year period, Region 2 had the largest number of construction completions at nonfederal NPL sites, 104, followed by Region 5 with 95 sites.

According to EPA officials, one of the reasons for the decrease in the number of construction completions was the decline from fiscal years 1999 through 2013 in the total number of nonfederal sites that were available for construction completion. Our analysis of EPA data indicates that, while the number of sites available for construction completion has declined, so too has the number of construction completions compared to those sites available for construction completion as shown in figure 14. For example, in fiscal year 1999, there were 80 construction completions at nonfederal NPL sites out of 630 available for construction completion (or about 13 percent). However, in fiscal year 2013, there were 13 construction completions out of 428 (or about 3 percent).

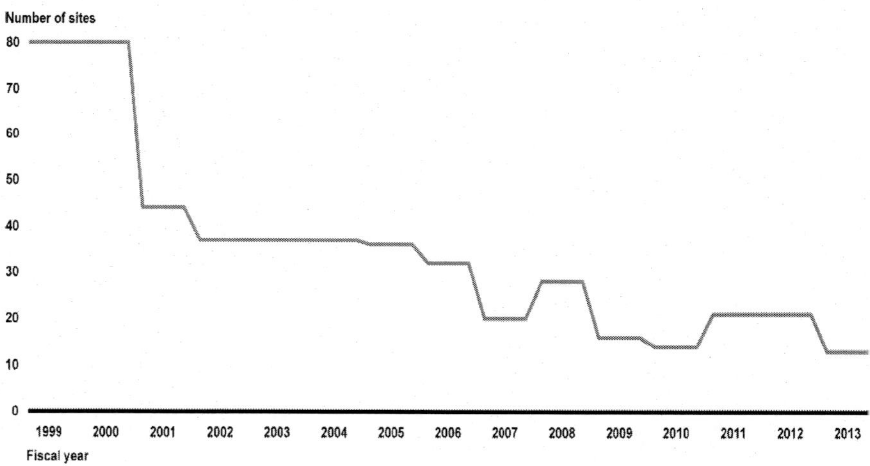

Source: GAO analysis of EPA data. | GAO-15-812

Figure 13. Number of Construction Completions at Nonfederal National Priorities List Sites, Fiscal Years 1999 through 2013.

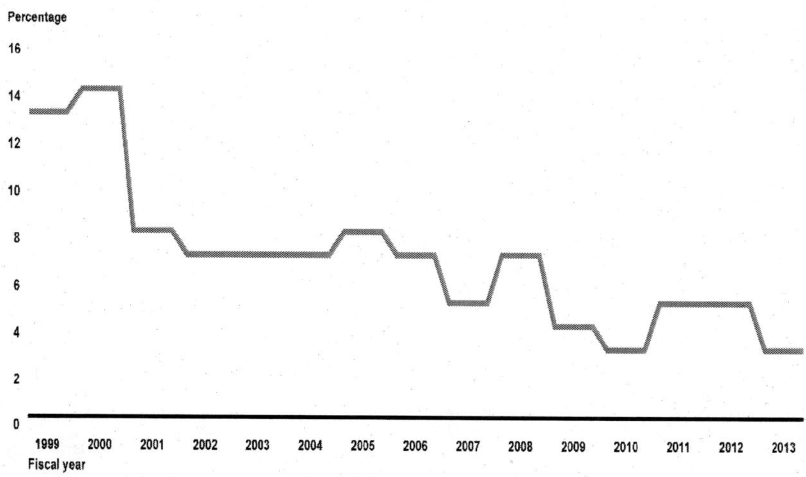

Source: GAO analysis of EPA data. | GAO-15-812

Figure 14. Percentage of Construction Completions at Nonfederal National Priorities List Sites Out of the Total Number of Nonfederal National Priorities List Sites That Were Available for Construction Completion, Fiscal Years 1999 through 2013.

AGENCY COMMENTS

We requested comments on a draft of this product from EPA. EPA did not provide written comments. In an e-mail received on September 11, 2015, the Audit Liaison stated that EPA agreed with our report's findings and provided technical comments. We incorporated these technical comments, as appropriate.

J. Alfredo Gómez
Director,
Natural Resources and Environment

APPENDIX I: OBJECTIVES, SCOPE, AND METHODOLOGY

This appendix provides information on the objectives, scope of work, and the methodology used to determine, for fiscal years 1999 through 2013, the trends in (1) the annual federal appropriations to the Superfund program and

Environmental Protection Agency (EPA) expenditures for remedial cleanup activities at nonfederal sites on the National Priorities List (NPL) and (2) the number of nonfederal sites on the NPL, the number of remedial action project completions, and the number of construction completions at nonfederal NPL sites.

To determine the trend in the annual federal appropriations to the Superfund program and EPA expenditures for remedial cleanup activities at nonfederal sites on the NPL from fiscal years 1999 through 2013, we reviewed and analyzed Superfund program funding data. In addition, we analyzed expenditure data from EPA's Integrated Financial Management System for fiscal years 1999 through 2003, and from its replacement financial system Compass, for fiscal years 2004 through 2013.[1] These data included Superfund agency expenditures from annual appropriations, including American Recovery and Reinvestment Act of 2009 funds, but they excluded expenditures of Homeland Security Supplemental appropriation, special accounts, and state cost share funds, as well as funds received from other agencies (i.e., funds-in interagency agreements and intergovernmental personnel agreements) and expenditures in support of Brownfields program activities. EPA provided agencywide data for site and nonsite expenditures segregated by expenditure category and source of funding. EPA provided the financial data in nominal values, which we converted to constant 2013 dollars. We analyzed these data to identify the trend in total expenditures of annual federal appropriations for, among other things, the remedial action cleanup process and the median expenditure by site and type of site (e.g., mining and manufacturing). The scope of our analyses for both objectives varied from year-to-year because we examined only nonfederal sites that were "active," i.e., on the NPL at any given point during the fiscal year. We also obtained and analyzed information on the nonfederal NPL sites that, according to EPA, had remedial action projects that were ready to begin but were not funded because of resource constraints.

To determine the trend in the number of nonfederal sites on the NPL, the number of remedial action project completions, and the number of construction completions at nonfederal NPL sites from fiscal years 1999 through 2013, we analyzed EPA's program data from fiscal years 1999 through 2013. At the time of our analysis, EPA officials stated that 2013 would be the most recent year with complete and stable data, and these data were available in the agency's Comprehensive Environmental Response, Compensation, and Liability Information System (CERCLIS) database. As of June 2015, EPA officials stated that the agency was not in a position to release data for fiscal year 2014 that would be comparable to the fiscal years 1999 through 2013 data until fiscal year 2016. However, in July 2015, EPA officials were able to provide fiscal year 2014 data on the number of new nonfederal sites added to the NPL, nonfederal sites deleted from the NPL, remedial action project completions, and construction completions because the agency gathers these data through manual data requests for which each EPA regional office certifies the data that it provides to EPA Headquarters. We obtained data from EPA for all of the nonfederal sites that were or had been on the NPL, as of the end of fiscal year 2013.[2] One site, the Ringwood Mines/Landfill site, had two final dates—the date a site is formally added to the NPL via a *Federal Register* notice—because the site was restored to the NPL after it had been deleted. We used the latest final date that was provided by EPA in our analysis. The Ringwood Mine/Landfill site was included in the results of our analysis of new nonfederal sites added to the NPL and number of nonfederal sites on the NPL, but we excluded it from our analysis of the median amount of time between when a site is proposed and when it is added to the NPL. Our analysis included nonfederal sites that were on the NPL, including sites that had been deleted, through fiscal year 2013. We analyzed site-level data for nonfederal NPL sites to summarize trends in the number of new nonfederal sites added to the NPL and the number of nonfederal sites that reached construction completion and deletion. We also analyzed the number of remedial action project completions in each of the 15 years in our analysis. Our analysis did not include (1) four sites that started off on the NPL but were deferred to another authority and deleted from the NPL and (2) five sites that were proposed but never became final on the NPL.

To address both objectives, we reviewed agency documents including, for example, the *Superfund Program Implementation Manual*, and we interviewed EPA officials in headquarters and Region 2 to discuss the trends we identified in our analyses and potential reasons for these trends. We spoke with EPA staff in Region 2 because Region 2 sites received the most site-specific cleanup funds for remedial cleanup activities, Region 2 had the state—New York—with the largest population living within a 3-mile buffer of its nonfederal NPL sites, as of fiscal year 2013, and included the state—New Jersey—that had the largest number of nonfederal NPL sites in fiscal year 2013. We also interviewed knowledgeable stakeholders from the Association of State and Territorial Solid Waste Management Officials and the National Academy of Sciences. Additionally, we reviewed prior GAO reports on EPA's Superfund program. A list of related GAO products is included at the end of this report.

To assess the reliability of the data from the EPA databases used in this report, we reviewed relevant documents, such as the 2013 CERCLIS data entry control plan guidance and regions' CERCLIS data entry control plans; examined the data to identify obvious errors or inconsistencies; compared the data that we received to publicly available data; and interviewed EPA officials. We determined the data to be sufficiently reliable for the purposes of this report.

In addition, to determine the estimated population that lived within 3 miles of nonfederal sites on the NPL, we generally relied on EPA's Office of Solid Waste and Emergency Response methodology and analyzed data from (1) CERCLIS on the 1,158 nonfederal sites on the NPL in the 50 states and U.S. territories (Guam, Puerto Rico, and the Virgin Islands), as of the end of fiscal year 2013, and (2) Census from the 2009 through 2013 American Community Survey 5-year estimate[3] for the 1,141 nonfederal sites in the 50 states and the District of Columbia. A circular site boundary, equal to the site acreage, was modeled around the latitude/longitude for each site and then a 3-mile buffer ring was placed around the site boundary. For the 138 sites in 34 states that EPA did not have acreage information, a circular site boundary was modeled around the latitude/longitude point, and then a 3-mile buffer ring was placed around the point. American Community Survey data was then collected for each block group with a centroid that fell within the 3-mile area and rounded. Percentage numbers were rounded to the nearest whole percent.

We conducted this performance audit from October 2014 to September 2015 in accordance with generally accepted government auditing standards. Those standards require that we plan and perform the audit to obtain sufficient, appropriate evidence to provide a reasonable basis for our findings and conclusions based on our audit objectives. We believe that the evidence obtained provides a reasonable basis for our findings based on our audit objectives.

Appendix II: Estimated Population, by State, That Lived within 3 Miles of a Nonfederal Site on the National Priorities List as of Fiscal Year 2013

State	Number of nonfederal NPL sites	Total estimated state population (thousands)	Estimated state population that lived within 3 miles of a nonfederal NPL site (thousands)	Percentage of total estimated state population that lived within 3 miles of a nonfederal NPL site	Estimated state population under the age of 18 that lived within 3 miles of a nonfederal NPL site (thousands)	Estimated state population 65 years and older that lived within 3 miles of a nonfederal NPL site (thousands)
New York	83	19,487	5,579	29	1,162	736
California	74	37,659	5,309	14	1,299	604
New Jersey	107	8,832	4,454	50	1,037	568
Florida	49	19,091	2,686	14	579	396
Pennsylvania	89	12,731	2,529	20	557	369
Washington	36	6,820	1,755	26	399	205
Texas	46	25,639	1,712	7	457	155
Michigan	65	9,886	1,330	14	314	167
Minnesota	23	5,348	1,169	22	248	136
Massachusetts	25	6,605	999	15	224	140
Indiana	34	6,515	993	15	231	126
Ohio	34	11,550	850	7	200	123
Illinois	40	12,849	741	6	197	90
Wisconsin	38	5,707	707	12	165	99
Missouri	30	6,007	651	11	157	87
Colorado	15	5,119	621	12	143	58
North Carolina	35	9,651	620	6	144	73

(Continued)

State	Number of nonfederal NPL sites	Total estimated state population (thousands)	Estimated state population that lived within 3 miles of a nonfederal NPL site (thousands)	Percentage of total estimated state population that lived within 3 miles of a nonfederal NPL site	Estimated state population under the age of 18 that lived within 3 miles of a nonfederal NPL site (thousands)	Estimated state population 65 years and older that lived within 3 miles of a nonfederal NPL site (thousands)
Utah	11	2,814	546	19	142	56
Virginia	20	8,101	477	6	116	54
Connecticut	13	3,584	463	13	105	66
Maryland	10	5,834	398	7	91	48
Iowa	10	3,063	363	12	82	50
Arizona	7	6,480	361	6	91	35
Nebraska	12	1,842	346	19	90	39
Delaware	12	908	323	36	77	39
South Carolina	24	4,680	320	7	73	41
Oregon	12	3,869	302	8	72	35
Rhode Island	10	1,052	280	27	60	40
New Hampshire	19	1,319	253	19	58	32
New Mexico	13	2,070	251	12	58	34
Georgia	14	9,810	232	2	60	30
Montana	16	999	206	21	43	29
Louisiana	8	4,568	200	4	47	25
Kansas	11	2,868	194	7	54	20
Tennessee	13	6,402	176	3	41	22

State	Number of nonfederal NPL sites	Total estimated state population (thousands)	Estimated state population that lived within 3 miles of a nonfederal NPL site (thousands)	Percentage of total estimated state population that lived within 3 miles of a nonfederal NPL site	Estimated state population under the age of 18 that lived within 3 miles of a nonfederal NPL site (thousands)	Estimated state population 65 years and older that lived within 3 miles of a nonfederal NPL site (thousands)
Mississippi	8	2,977	129	4	33	17
West Virginia	7	1,854	119	6	21	17
Vermont	11	626	103	16	18	14
Arkansas	9	2,933	80	3	21	11
Idaho	4	1,583	79	5	21	10
Kentucky	13	4,361	74	2	17	12
Hawaii	1	1,376	71	5	19	7
Alabama	11	4,799	63	1	16	8
Maine	10	1,328	52	4	11	7
Oklahoma	5	3,786	46	1	12	7
Nevada	1	2,730	22	1	5	3
Wyoming	1	570	15	3	4	2
South Dakota	1	825	1	0	0	0
Alaska	1	720	.	.	.	.
District of Columbia	0	619	.	.	.	.

(Continued)

State	Number of nonfederal NPL sites	Total estimated state population (thousands)	Estimated state population that lived within 3 miles of a nonfederal NPL site (thousands)	Percentage of total estimated state population that lived within 3 miles of a nonfederal NPL site	Estimated state population under the age of 18 that lived within 3 miles of a nonfederal NPL site (thousands)	Estimated state population 65 years and older that lived within 3 miles of a nonfederal NPL site (thousands)
North Dakota	0	690	.	.	.	.
Guam	1	Not available				
Puerto Rico	15	Not available	.	.	.	.
Virgin Islands	1	Not available	.	.	.	.

Source: GAO analysis of EPA data and U.S. Census data. | GAO-15-812

Note: The methodology for GAO's analysis is generally based on EPA's Office of Solid Waste and Emergency Response's approach. Data analyzed include (1) 1,158 nonfederal sites on the National Priorities List (NPL), in the 50 states and U.S. territories (Guam, Puerto Rico, and the Virgin Islands), as of the end of fiscal year 2013 and (2) Census data from the 2009-2013 American Community Survey 5-year estimate for the 1,141 nonfederal NPL sites in the 50 states and the District of Columbia. A circular site boundary, equal to the site acreage, was modeled around the latitude/longitude for each site and then a 3-mile buffer ring was placed around the site boundary. For the 138 sites in 34 states that EPA did not have acreage information, a circular site boundary was modeled around the latitude/longitude point, and then a 3-mile buffer ring was placed around the point. American Community Survey data was then collected for each block group with a centroid that fell within the 3-mile area and rounded to the nearest 1,000. Percentage numbers were rounded to the nearest whole percent.

APPENDIX III: NONFEDERAL SITES ON THE NATIONAL PRIORITIES LIST, FISCAL YEARS 1999 THROUGH 2013

Appendix III provides information from fiscal years 1999 through 2013 on the number of nonfederal sites on the National Priorities List (NPL) at the beginning and end of the fiscal year after accounting for new sites added to and existing sites deleted from the NPL during the fiscal year (table 2); and the number of nonfederal sites on the NPL by site type for each fiscal year (table 3).

Table 2. Nonfederal Sites on the National Priorities List, Fiscal Years 1999 through 2013

Fiscal year	Number of nonfederalsites on the NPL at thestart of the fiscal year	New nonfederal sites added to the NPL	Nonfederal sites deleted from the NPL	Number of nonfederalsites on the NPL at theend of the fiscal year
1999	1,039	37	22	1,054
2000	1,054	36	18	1,072
2001	1,072	28	30	1,070
2002	1,070	18	14	1,074
2003	1,074	20	9	1,085
2004	1,085	11	16	1,080
2005	1,080	17	17	1,080
2006	1,080	11	7	1,084a
2007	1,084	12	6	1,090a
2008	1,090	18	9	1,099a
2009	1,099	19	8	1,110a
2010	1,110	20	7	1,123a
2011	1,123	25	7	1,141a
2012	1,141	24	9	1,156a
2013	1,156	8	6	1,158a

Source: GAO analysis of EPA data. | GAO-15-812.

[a] The total includes one site—the Ringwood Mines/Landfill—that was deleted from the National Priorities List (NPL) in 1994 and restored to the NPL in 2006.

Table 3. Number of Nonfederal Sites on the National Priorities List, by Site Type, Fiscal Years 1999 through 2013

Fiscal year	Site type						
	Manufacturing	Mining	"Multiple"	"Other"	Recycling	Waste management	Total
1999	390	10	31	121	93	409	**1,054**
2000	399	13	31	127	95	407	**1,072**
2001	401	18	31	132	93	395	**1,070**
2002	402	20	32	136	94	390	**1,074**
2003	405	23	31	141	96	389	**1,085**
2004	405	25	31	145	96	378	**1,080**
2005	408	27	31	149	94	371	**1,080**
2006	411	28	31	152	92	370	**1,084a**
2007	417	29	30	153	93	368	**1,090a**
2008	419	33	30	159	92	366	**1,099a**
2009	423	37	31	164	92	363	**1,110a**
2010	426	39	31	173	92	362	**1,123a**
2011	435	43	31	180	92	360	**1,141a**
2012	443	44	31	186	94	358	**1,156a**
2013	448	44	30	188	94	354	**1,158a**

Source: GAO analysis of EPA data. | GAO-15-812.

Note: The "multiple" site type includes sites with operations that fall into more than one of EPA's categories. The "other" site type includes sites that often have contaminated sediments or groundwater plumes with no identifiable source.

[a] The total includes one site—the Ringwood Mines/Landfill—that was deleted from the National Priorities List (NPL) in 1994 and restored to the NPL in 2006.

End Notes

[1] Pub. L. No. 96-510, 94 Stat. 2767 (1980) (codified as amended at 42 U.S.C. §§ 9601 – 9675 (2015)).

[2] Federal facilities are sites owned or operated by a department, agency, or instrumentality of the United States, such as the Departments of Defense, Energy, and the Interior. These agencies may have a significant role in the cleanup of these facilities, and such cleanups are funded by the agency and not by EPA's Superfund appropriation. Processes and provisions specific to these federal sites are generally not discussed in this report.

[3] Unless otherwise indicated, all dollars and percentage calculations are in constant 2013 dollars.

[4] Polychlorinated biphenyls belong to a broad family of man-made organic chemicals known as chlorinated hydrocarbons which are chemical compounds of chlorine, hydrogen, and carbon atoms only.

[5] GAO, *Superfund: EPA's Estimated Costs to Remediate Existing Sites Exceed Current Funding Levels, and More Sites Are Expected to Be Added to the National Priorities List*, GAO-10-380 (Washington D.C.: May 6, 2010); *Superfund: Litigation Has Decreased and EPA Needs Better Information on Site Cleanup and Cost Issues to Estimate Future Program Funding Requirements*, GAO-09-656 (Washington D.C.: July 15, 2009).

[6] At the time of our review, fiscal year 2013 was the most recent year with complete and stable program data, according to EPA officials. We used expenditure data that were comparable with the same timeframe for which program data were available.

[7] Under CERCLA, PRPs generally include current or former owners or operators of a site or the generators and transporters of the hazardous substances.

[8] According to EPA guidance, EPA uses operable units and remedial action projects to subdivide a Superfund site into a series of smaller components that allow for effective management and implementation of cleanup activities. An operable unit is a discrete action that comprises an incremental step in cleaning up a site and commonly refers to a geographic area, a pathway of the contamination (e.g., groundwater), or type of remedy. A site may consist of one or more operable units, each of which may be addressed by one or more remedial action projects. A remedial action project is generally where the physical work undertaken to address contamination takes place at a site.

[9] 65 Fed. Reg. 57,810 (Sept. 26, 2000).

[10] There is no requirement that PRPs maintain or disclose their cleanup costs, and they generally consider such cost information to be confidential. According to EPA officials, the agency relies on the estimated remedy construction and maintenance costs identified in the record of decision to estimate the value of cleanups conducted by PRPs.

[11] Census data is from the 2009-2013 American Community Survey 5-year estimate for the 50 states and the District of Columbia.

[12] Annual federal appropriations declined from about $1.5 billion to about $1.1 billion—over 26 percent—in nominal dollars. The annual appropriation to the Superfund program for fiscal year 2014 was about $1.1 billion.

[13] The Recovery Act was enacted with the purpose to promote economic recovery, make investments, and minimize and avoid reductions in state and local government services, among other things. Pub. L. No. 111-5, 123 Stat. 115.

[14] Recovery Act funds were $600 million in nominal dollars. Of the $600 million, EPA allocated $582 million to remedial cleanup activities and $18 million to internal EPA activities related to the management, oversight, and reporting of Recovery Act funds.

[15] Site-specific cleanup fund expenditures from annual appropriations declined from about $0.5 billion to $0.4 billion—about 18 percent—in nominal dollars.

[16] According to EPA officials, EPA obligated Recovery Act funds in fiscal year 2009 but, according to our analysis of EPA data, the majority of the funds were spent through fiscal year 2011.

[17] The Office of Solid Waste and Emergency Response manages the Superfund program.

[18] According to EPA officials, there are some large and relatively costly nonfederal NPL sites that require annual site-specific funding agreements. These are agreements between EPA headquarters and regional offices that provide a planned funding amount that a site will receive in each future fiscal year. These funding agreements are determined after the ranking of a project by the Panel, and approval is given to begin work. In general, this

occurs for sites where remediation is anticipated to continue over multiple fiscal years and is expected to cost more than $100 million.

[19] This amount is in nominal dollars.

[20] According to EPA officials, in fiscal year 2014, EPA did not fund five new remedial action projects out of a total of 31.

[21] According to an EPA official, deobligated funds cannot be readily identified in the financial database before fiscal year 2003, because EPA had not established the specific fund code for deobligations until fiscal year 2002.

[22] EPA consistently spent the most nonfederal NPL cleanup funds per year from fiscal years 1999 through 2013 in Region 2. On average, EPA spent about $165 million per year in Region 2, followed by about $47 million in Region 8, and about $46 million in Region 1.

[23] EPA spent the most cleanup funds in every year from fiscal years 1999 through 2013 on New Jersey sites, according to our analysis of EPA data.

[24] Cleanup work performed directly by PRPs is not included in our analysis because EPA does not track cost data on PRP-led projects

[25] Throughout the report, we used the median value due to the large variance in the per-site EPA financial and program data.

[26] The median per-site annual expenditures declined by about 30 percent or from about $27,400 to about $19,100 in nominal dollars.

[27] An outyear is any fiscal year beyond the budget year for which projections are made in, for example, the President's budget submission.

[28] More information on the funding of the Bunker Hill cleanup appears at http://yosemite.epa.gov/R10/CLEANUP.NSF/7780249BE8F251538825650F0070BD8B/W ho+pays+for+the+Bunker+Hill+Superfund+Site+cleanup (accessed July 9, 2015).

[29] Our analysis of EPA data included all remedial action project completions and construction completions at the nonfederal NPL sites from fiscal years 1999 through 2013, regardless of who performed the activity.

[30] According to our 2013 report on alternatives to placing sites on the NPL, Massachusetts, New Jersey, and California were among the states with the most mature environmental programs and had 247 sites, 221 sites, and 180 sites, respectively, in their states' program. For more information, see GAO, *Superfund: EPA Should Take Steps to Improve Its Management of Alternatives to Placing Sites on the National Priorities List*, GAO-13-252 (Washington, D.C.: Apr. 9, 2013).

[31] Owners or operators of active facilities that treat, store, or dispose of hazardous waste must take corrective actions to clean up contamination from the facility under the Resource Conservation and Recovery Act. If the owner or operator goes bankrupt, they may be unable to complete the corrective action. If the site is referred to the Superfund program, and subsequently added to the NPL, then federal funding may be used to complete the cleanup.

[32] According to EPA officials, 21 new nonfederal sites were added to the NPL in fiscal year 2014.

[33] This calculation excludes the Ringwood Mines/Landfill site located in New Jersey, which was deleted from the NPL in 1994 and restored to the NPL in 2006.

[34] According to EPA officials, EPA's goal is to clean up a site. As such, the agency will often hold off on making a listing decision to allow time for negotiations or cleanup to progress.

[35] According to EPA officials, owners of recycling and dry cleaning sites generally do not have the assets to clean up the site, and dry cleaning sites generally have groundwater contamination, which is expensive to clean up.

[36] According to EPA officials, EPA deleted 14 nonfederal sites from the NPL in fiscal year 2014.

[37] According to EPA officials, for fiscal year 2014, EPA added 21 nonfederal sites to the NPL and deleted 14, resulting in a total of 1,165 sites on the NPL.

[38] The decline in the number of remedial action project completions continued into fiscal year 2014 with 61 completions, according to EPA officials.

[39] The decline in the number of construction completions at nonfederal NPL sites continued into fiscal year 2014, with 7, according to EPA officials.

End Notes for Appendix I

[1] At the time of our review, fiscal year 2013 was the most recent year with complete and stable program data, according to EPA officials. We used expenditure data that were comparable with the same timeframe for which program data were available.

[2] These data excluded any sites that were proposed, removed, or withdrawn from the NPL.

[3] The American Community Survey is an ongoing survey on topics such as social, economic, demographic, and housing characteristics of the U.S. population. The 5-year estimates from the American Community Survey are "period" estimates that represent data collected over a period of time. The primary advantage of using multiyear estimates is the increased statistical reliability of the data for less populated areas and small population subgroups. The most recent 5-year estimate covers 2009 through 2013. Because the American Community Survey data are based on probability samples, estimates are formed using the appropriate estimation weights provided with each survey's data. Because each of these samples follows a probability procedure based on random selection, they represent only one of a large number of samples that could have been drawn. Since each sample could have provided different estimates, we express our confidence in the precision of our particular sample's results as a percentage of the estimate, the sampling error divided by the estimate. Unless otherwise noted, all estimates have errors of 5 percent or less.

In: EPA Cleanup Approaches …
Editor: Allan R. Payne

ISBN: 978-1-63485-260-9
© 2016 Nova Science Publishers, Inc.

Chapter 2

SUPERFUND: EPA SHOULD TAKE STEPS TO IMPROVE ITS MANAGEMENT OF ALTERNATIVES TO PLACING SITES ON THE NATIONAL PRIORITIES LIST[*]

United States Government Accountability Office

WHY GAO DID THIS STUDY

Under the Superfund program, EPA may address the long-term cleanup of certain hazardous waste sites by placing them on the National Priorities List (NPL) and overseeing the cleanup. To be eligible for the NPL, a site must be sufficiently contaminated, among other things. EPA regions have discretion to choose among several other approaches to address sites eligible for the NPL. For example, under the Superfund program, EPA regions may enter into agreements with Potentially Responsible Parties (PRPs) using the Superfund Alternative (SA) approach. EPA may also defer the oversight of cleanup at eligible sites to approaches outside of the Superfund program. GAO was asked to review EPA's implementation of the SA approach and how it compares with the NPL approach. This report examines (1) how EPA addresses the cleanup of sites it has identified as eligible for the NPL, (2) how the processes for implementing the SA and NPL approaches compare, and (3) how SA

---

[*] This is an edited, reformatted and augmented version of the United States Government Accountability Office publication, GAO-13-252, dated April 2013.

agreement sites compare with similar NPL sites in completing the cleanup process. GAO reviewed applicable laws, regulations, and guidance; analyzed program data as of December 2012; interviewed EPA officials; and compared SA agreement sites with 74 NPL sites selected based on their similarity to SA agreement sites.

WHAT GAO RECOMMENDS

GAO recommends, among other things, that EPA issue guidance to define and clarify documentation requirements for OCA deferrals and clarify its policies on SA agreement sites. EPA agreed with the report's recommendations.

WHAT GAO FOUND

The Environmental Protection Agency (EPA) most commonly addresses the cleanup of sites it has identified as eligible for the National Priorities List (NPL) by deferring oversight of the cleanup to approaches outside of the Superfund program. As of December 2012, of the 3,402 sites EPA identified as potentially eligible, EPA has deferred oversight of 1,984 sites to approaches outside the Superfund program, including 1,766 Other Cleanup Activity (OCA) deferrals to states and other entities. However, EPA has not issued guidance for OCA deferrals as it has for the other cleanup approaches. Moreover, EPA's program guidance does not clearly define each type of OCA deferral or specify in detail the documentation EPA regions should have to support their decisions on OCA deferrals. Without clearer guidance on OCA deferrals, EPA cannot be reasonably assured that its regions are consistently tracking these sites or that their documentation will be appropriate or sufficient to verify that these sites have been deferred or have completed cleanup. Under the Superfund program, EPA oversees the cleanup of 1,313 sites on the NPL, 67 sites under the Superfund Alternative (SA) approach, and at least 38 sites under another undefined approach.

The processes for implementing the SA and NPL approaches, while similar in many ways, have several differences. EPA has accounted for some of these differences in its SA guidance by listing specific provisions for SA agreements with potentially responsible parties (PRP), such as owners and operators of a site. One such provision helps ensure cleanups are not delayed by a loss of funding if the PRP stops cleaning up the site. However, some EPA regions have entered into agreements with PRPs at sites that officials said were likely eligible for the SA approach without following the SA guidance. Such agreements may not benefit from EPA's provisions for SA agreements. EPA headquarters officials said the agency prefers regions to use the SA approach at such sites, but EPA has not stated this preference explicitly in its guidance. In addition, EPA's tracking and reporting of certain aspects of the process under the SA approach differs from that under the NPL approach. As a result, EPA's tracking of SA agreement sites in its Superfund database is incomplete; the standards for documenting the NPL eligibility of SA agreement sites are less clear than those for NPL sites; and EPA is not publicly reporting a full picture of SA agreement sites. Unless EPA makes improvements in these areas, its management of the process at SA agreement sites may be hampered.

The SA agreement sites showed mixed results in completing the cleanup process when compared with 74 similar NPL sites GAO analyzed. Specifically, SA agreement and NPL sites in GAO's analysis showed mixed results in the average time to complete negotiations with PRPs and for specific cleanup activities, such as remedial investigation and feasibility studies, remedial designs, and remedial actions. In addition, a lower proportion of SA agreement sites have completed cleanup compared with similar NPL sites. SA agreement sites tend to be in earlier phases of the cleanup process because the SA approach began more recently than the NPL approach. Given the limited number of activities for both NPL and SA agreement sites in GAO's analysis, these differences cannot be attributed entirely to the type of approach used at each site.

ABBREVIATIONS

CERCLA	Comprehensive Environmental Response, Compensation, and Liability Act
CERCLIS	Comprehensive Environmental Response, Compensation, and Liability Information System
EPA	Environmental Protection Agency
GPRA	Government Performance and Results Act of 1993
IG	Inspector General
NPL	National Priorities List
NRC	Nuclear Regulatory Commission

NRD	natural resource damages
OCA	Other Cleanup Activity
PRP	potentially responsible party
RCRA	Resource Conservation and Recovery Act
SA	Superfund Alternative

* * *

April 9, 2013

The Honorable Henry A. Waxman
Ranking Member
Committee on Energy and Commerce
House of Representatives

The Honorable John D. Dingell
House of Representatives

The Environmental Protection Agency (EPA) estimates that one in four Americans lives within 3 miles of a hazardous waste site. Many hazardous waste sites pose serious risks to human health and the environment, and their cleanup can be expensive and take many years to complete. While these sites may not necessarily be subject to a federal cleanup requirement, several approaches exist to address such long-term cleanups. EPA manages the Superfund program—the federal government's principal program to clean up hazardous waste sites— under the Comprehensive Environmental Response, Compensation, and Liability Act (CERCLA) of 1980.[1] Under this program, EPA can place sites with contamination that is sufficiently severe on the National Priorities List (NPL), a list of sites for attention under the federal Superfund program that includes sites among the nation's most seriously contaminated.[2] At sites on the NPL, EPA oversees the cleanup, which may be performed by potentially responsible parties (PRP)[3] or by EPA itself. Aside from placing an eligible site on the NPL, EPA can also ensure cleanup through the Superfund Alternative (SA) approach.[4] Under this approach, PRPs agree to clean up sites (hereafter referred to as SA agreement sites), and EPA does not list the sites on the NPL at that time, which allows PRPs to avoid the perceived stigma of having a site on the NPL. Since EPA first issued guidance on the SA approach in 2002, EPA and its Office of Inspector General (IG) have evaluated the approach and made various recommendations to improve its implementation. In 2010, EPA reported to the IG that it had implemented the recommended actions.[5] Where EPA decides not to address the site under the Superfund program (i.e., list the site on the NPL, use the SA approach, or otherwise retain oversight), EPA may defer[6] sites whose contamination makes them eligible for the NPL to other federal and state cleanup approaches.[7]

EPA's regional offices may discover potential hazardous waste sites, or such sites may come to EPA's attention through reports from other federal agencies, state agencies, or citizens. EPA then reviews available information about each site to decide whether to add it to the Superfund program database—the Comprehensive Environmental Response, Compensation, and Liability Information System (CERCLIS). At some point after a site is added to CERCLIS, EPA assesses the severity of the contamination at the site to determine whether the site is eligible to be placed on the NPL. If the site has some contamination but is not eligible for the NPL, EPA will not pursue a long-term cleanup under the Superfund program; state cleanup programs or other programs may still address the contamination at the site. Long-term cleanups at sites eligible for the NPL that are conducted under the Superfund program generally follow an established process. This process consists of several phases, including studying site conditions, selecting a method to clean up the site, and conducting the actual cleanup. In some cases, cleanup of a site is divided into smaller parts, known as operable units, and cleanup may proceed at different rates at each of these operable units.[8] To accomplish the cleanup, EPA may, among other options, negotiate and enter into agreements with PRPs for them to address contamination at the site. These agreements may cover one or more phases of the cleanup process and may address one or more operable units. Furthermore, EPA may enter into more than one agreement with a PRP at a site. Alternately, for sites listed on the NPL, EPA may conduct the cleanup itself and pursue costs from PRPs through administrative or judicial actions.

In addition to conducting long-term cleanups, EPA may use its Superfund emergency response authorities to conduct removal actions at sites. Removal actions are usually short-term cleanups at sites that pose immediate threats to human health or the environment. Under the removal program, EPA has conducted thousands of cleanup actions instead of or in combination with long-term cleanups.

In this context, you asked us to review the implementation of the SA approach and how this approach compares to listing sites on the NPL. Our objectives were to examine (1) how EPA addresses the cleanup of sites it has identified as eligible for the NPL, (2) how the processes for implementing the

SA and NPL approaches compare, and (3) how SA agreement sites compare with similar NPL sites in completing the cleanup process.

To conduct this work, we analyzed applicable federal laws and EPA regulations and guidance to understand the available approaches to address hazardous waste sites that are reported to the Superfund program and have a level of contamination that makes them eligible for the NPL. We also conducted interviews with officials in all 10 EPA regions. To determine how EPA addresses the cleanup of sites eligible for the NPL, we analyzed EPA data from the CERCLIS database to establish the number of such sites that are being addressed through each approach as of December 2012. To assess the reliability of these data, we analyzed related documentation, examined the data for errors or inconsistencies, and interviewed agency officials about any known data problems and to learn more about their procedures for maintaining the data. We determined the data to be sufficiently reliable for calculating durations for completing different cleanup activities, including negotiations, at SA agreement and NPL sites. We also interviewed a nonprobability, convenience sample of officials from 13 state cleanup programs who were familiar with available cleanup approaches. The sample consisted of representatives from state environmental departments taking part in an Association of State and Territorial Solid Waste Management Officials conference call who agreed to speak with us. Because this was a nonprobability sample, the results of our analysis cannot be generalized to all states; however, these officials provided important information about the cleanup process. To compare the processes for implementing the SA and NPL approaches, including the cleanup process and EPA's oversight, we discussed the processes with EPA headquarters officials and officials from all 10 of EPA's regions, obtained relevant supporting documentation, and analyzed previous reviews of the SA approach. To determine how SA agreement sites compare with similar NPL sites in completing the cleanup process, we constructed a comparison group of 74 NPL sites with agreements between EPA and PRPs similar to agreements at 66 SA agreement sites.[9] For example, all sites had agreements that were entered into from June 2002 through December 2012 that involved cleanup actions at the site. Additionally, to match the characteristics of most SA agreements, we included only NPL agreements involving relatively few PRPs and agreements whose estimated costs were similar to costs at SA agreement sites. Once we identified the group of NPL sites from these agreements, we compared the duration of relevant cleanup actions, including negotiations, with those at SA agreement sites. The 74 NPL sites that were selected for comparison with SA agreement sites are not representative of the universe of all NPL sites because they were selected based on specific criteria. For example, we selected for comparison only NPL sites at which a PRP agreed to conduct at least some part of the cleanup. The results of our analysis cannot be generalized to all NPL sites; however, these sites can provide important information about the cleanup process. Appendix I provides a more detailed description of our objectives, scope, and methodology.

We conducted this performance audit from December 2011 to April 2013 in accordance with generally accepted government auditing standards. Those standards require that we plan and perform the audit to obtain sufficient, appropriate evidence to provide a reasonable basis for our findings and conclusions based on our audit objectives. We believe that the evidence obtained provides a reasonable basis for our findings and conclusions based on our audit objectives.

BACKGROUND

This section discusses EPA's process for assessing sites under the Superfund program and the approaches identified by EPA for conducting long-term cleanups at sites eligible for the NPL under the Superfund program and under other available approaches.

EPA's Process for Assessing Sites under the Superfund Program

Under the Superfund program, EPA assesses hazardous waste sites for long-term cleanups through a specific process. At some point after a potential hazardous waste site is reported to the Superfund program and entered into CERCLIS, EPA regional officials, their contractors, or states acting under cooperative agreements with EPA evaluate the relative potential for a site to pose a threat to human health and the environment. EPA's 10 regional offices each are responsible for implementing Superfund within several states and, in some cases, territories. Under CERCLA, EPA may only pay for a remedial action at a site if the relevant state agrees, among other things, to pay a portion of the cleanup expenses, as well as all operations and maintenance costs. In addition, under a cooperative agreement with EPA, a state may assume the lead oversight role at a site in the Superfund program. Figure 1 shows the states included in each of the 10 EPA regions.

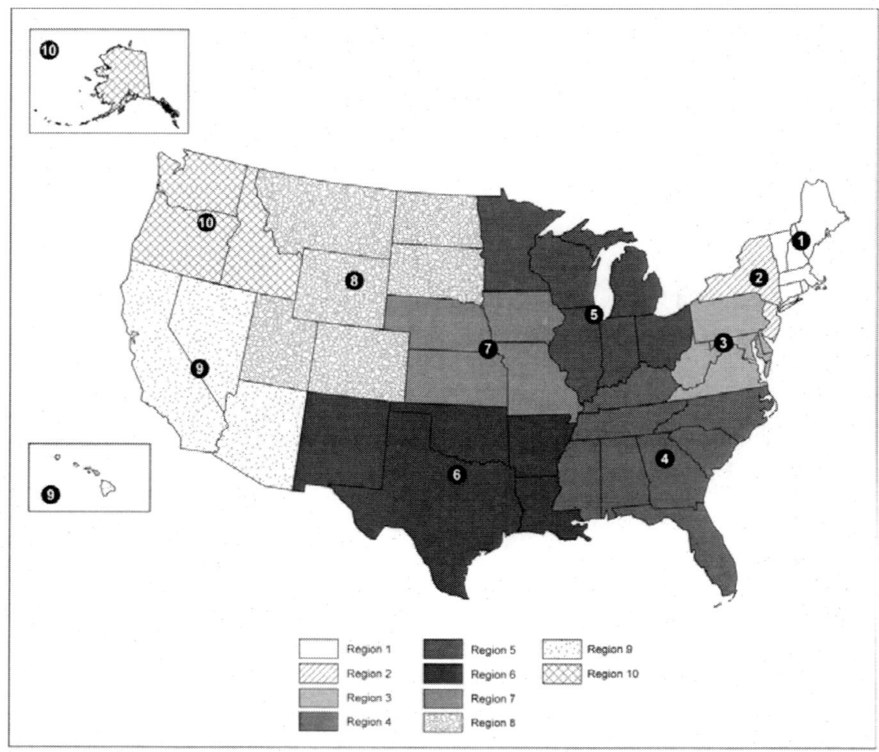

Sources: GAO analysis of EPA information; Map Resources (map).

Figure 1. EPA's 10 Regions.

During the initial phases of the long-term cleanup process—known as preliminary assessment and site inspection—EPA regional officials or their counterparts evaluate the potential need for additional investigation or action in connection with a release of hazardous substances from a site. Specifically, the preliminary assessment phase involves an evaluation of readily available information about a site and its surrounding area to determine if the release or potential release poses enough of a threat to human health and the environment that further investigation is needed. If further investigation is needed, a site inspection is performed. During this phase, investigators typically collect environmental and waste samples to determine what hazardous substances are present. Information collected during the preliminary assessment and site inspection is used to calculate and document a site's preliminary Hazard Ranking System score, which indicates a site's relative threat to human health and the environment based on potential pathways of contamination.[10] Sites with a Hazard Ranking System score of 28.50 or greater are eligible for listing on the NPL. Information collected from the initial assessment phases to develop Hazard Ranking System scores is not intended to be sufficient to determine either the extent of contamination or how to clean up a particular site. After a site is determined to be eligible for the NPL, EPA chooses which long-term cleanup approach is best suited to the site. In some cases, EPA may conduct a short-term cleanup known as a removal action or otherwise delay selection of a long-term cleanup approach.

Approaches Identified by EPA for Conducting Long-Term Cleanups at Sites Eligible for the NPL

EPA may choose among several approaches to address sites with a relative threat to human health and the environment that is sufficiently severe to make them eligible for listing on the NPL. For long-term cleanups, EPA can retain oversight of sites under the Superfund program or defer the oversight of sites to other approaches, as shown in figure 2.

Superfund Program Approaches for Conducting Long-Term Cleanups

Under its Superfund program, EPA conducts long-term cleanups using three approaches. The first and most common approach under the Superfund program involves listing a site on the NPL. To do so, EPA first proposes the site for listing on the NPL in the *Federal Register*. EPA then accepts public comments on the proposal and responds to the comments in a second and final *Federal Register* listing announcement of the site; then the agency may place on the NPL those sites that continue to meet the requirements for listing. The second approach that EPA may use under the Superfund program is the SA approach, which began informally in the 1990s whereby some EPA regions negotiated site cleanup agreements with PRPs for sites that PRPs, states, or local government officials and communities did not want to have listed on the NPL. To promote consistency across regions, EPA issued guidance in 2002 formalizing the SA approach, which it subsequently updated in 2004 and 2012.[11] According to EPA's guidance, to qualify for the SA approach, (1) a site's contamination must make it eligible for listing on the NPL; (2) EPA must anticipate a long-term cleanup at the site; and (3) there must be a willing, capable PRP who will negotiate and sign an agreement with EPA to perform the investigation or cleanup. The third approach EPA can use for long-term cleanup of sites is to address sites under the Superfund program but not list them on the NPL or address them through the SA approach.[12] These "Other" sites under the Superfund program can vary widely and include, among others, some sites with cleanup agreements that preceded the SA approach. Under these older agreements, for which there was no guidance at the time they were negotiated, EPA agreed not to list the site on the NPL, and the PRPs agreed to conduct the cleanup, according to EPA officials.

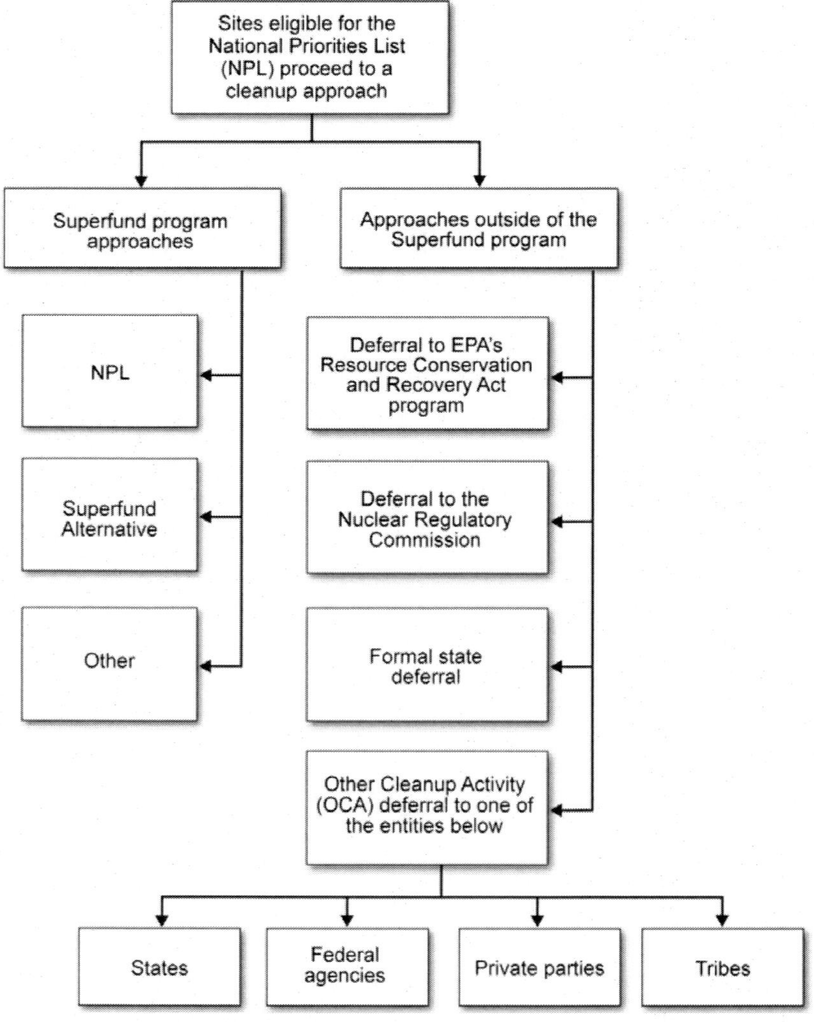

Source: GAO analysis of EPA information.

Figure 2. Approaches Identified by EPA to Clean Up Sites Eligible for the NPL.

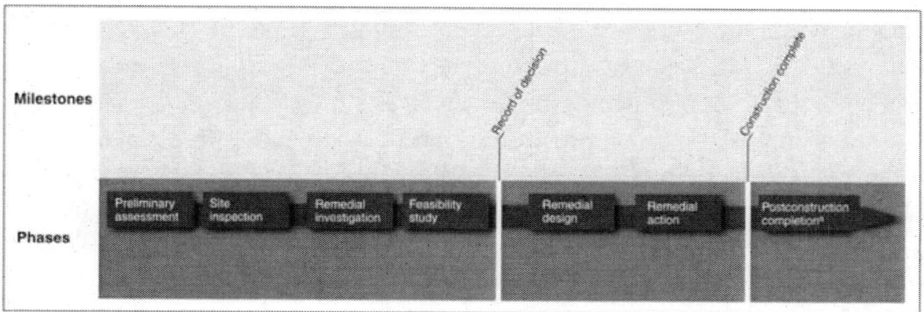

Source: GAO analysis of EPA information.

Note: Phases of the long-term cleanup process may overlap, and multiple phases may be concurrently under way at a site.

^a Postconstruction completion includes activities such as operation and maintenance, long-term response actions, and 5-year reviews, which ensure that Superfund cleanup actions provide for the long-term protection of human health and the environment.

Figure 3. Long-Term Cleanup Process at Sites under the Superfund Program.

Irrespective of the approach chosen, all sites under the Superfund program approaches follow the same general phases for long-term cleanup, as shown in figure 3, and EPA officials oversee the cleanup at all of these Superfund sites.

After the initial phases of the long-term cleanup process, EPA or a PRP conducts a two-part study of the site: (1) a remedial investigation to further characterize site conditions and assess the risks to human health and the environment, among other actions, and (2) a feasibility study to evaluate various cleanup options to address the problems identified in the remedial investigation. At the conclusion of these studies, EPA selects a remedy for addressing the site's contamination and develops a cost estimate for implementing the remedy; both of these are included in a record of decision. According to EPA officials, the level of cleanup depends on site-specific conditions, not the particular approach selected. EPA or a PRP then develops the method of implementation for the selected remedy during the remedial design phase and implements it during the remedial action phase, when actual cleanup of the site occurs. Multiple cleanup activities can occur within a given phase at the same or different operable units at one site. For example, one remedial action at an operable unit may address soil contamination, while another remedial action at the same operable unit may address groundwater contamination. When EPA or a PRP finishes the cleanup remedy at a site, all immediate threats have been addressed, and all long-term threats are under control, EPA generally considers the site to be "construction complete." For sites listed on the NPL, when EPA, in consultation with the state, determines that no further site response is appropriate, the agency may delete the site from the NPL. EPA reports achievements at NPL sites, including completion of some phases of the cleanup process, as part of the agency's implementation of provisions under the Government Performance and Results Act of 1993 (GPRA).[13] The act requires federal agencies to develop strategic plans with outcome oriented agency goals and objectives, performance measures to track the progress made toward achieving goals, annual goals linked to achieving the long-term goals, and annual reports on the results achieved. EPA does not report publicly the same achievements for sites that are not on the NPL.

CERCLA provides EPA with authority to enter into agreements with PRPs to conduct cleanup actions at sites;[14] this authority is relevant to all three approaches where EPA maintains oversight under the Superfund program. Model agreements for different phases of cleanup with standard provisions are to serve as the basis for negotiations and for the agency's legal documents. According to EPA officials, the agency typically uses legal documents known as "administrative orders on consent"—which do not require court approval— to record the agreements between EPA and PRPs for conducting remedial investigation and feasibility studies. An EPA agreement with a private party for conducting a remedial action generally takes the form of a "consent decree," which must be approved by a court.[15] Although agreements under the SA approach generally follow these model agreements, EPA guidance states that SA agreements are to include specific provisions, depending on the phase of cleanup to which the agreement applies. These provisions are intended to ensure equivalency between the SA and NPL approaches. In addition to negotiating agreements, EPA has authority to issue enforcement orders, such as "unilateral administrative orders," or to coordinate with the Department of Justice in seeking an injunction to require PRPs to conduct cleanup.[16]

Approaches Outside of the Superfund Program for Conducting Long-Term Cleanups

As an alternative to addressing a site under the Superfund program, EPA may defer oversight of the cleanup of a site eligible for the NPL to other cleanup approaches, including federal and state cleanup programs. For example, EPA may defer a site from its Superfund program to its Resource Conservation and Recovery Act (RCRA) program. Congress passed RCRA in 1976, establishing requirements, as well as giving EPA regulatory authority, for the generation, transportation, treatment, storage, and disposal of hazardous waste.[17] While CERCLA focuses on cleanup of sites where hazardous substances have been released, including inactive and abandoned hazardous waste sites, RCRA generally focuses on facilities currently generating, treating, storing, and disposing of hazardous waste—hazardous materials that are destined for disposal or recycling. RCRA authorizes EPA to issue administrative cleanup orders where an imminent and substantial danger to health and the environment may exist. At a given site, certain authorities of CERCLA and RCRA may be applicable to a cleanup. EPA also may defer certain sites to the Nuclear Regulatory Commission (NRC). The NRC licenses commercial nuclear facilities, including power reactors, and regulates and oversees their safe operation, including the decommissioning and decontamination associated with shutting down a licensed reactor. In what are known as "formal state deferrals," EPA may also defer sites to states or other entities, such as federally recognized tribes where applicable conditions are met. According to EPA guidance on this approach, the EPA region and the state cleanup program should enter into a memorandum of agreement certifying that the state has the necessary authority and capability to adequately supervise the PRP's cleanup actions, among other things. The state will then oversee cleanup actions conducted and funded by the PRPs at the site. The quality of these cleanup actions should be substantially similar to a cleanup required under CERCLA authorities, according to EPA guidance on this approach. Under this formal state deferral approach, the EPA region still negotiates the level of oversight appropriate for the particular site.

EPA may also defer oversight of the long-term cleanup of a site eligible for the NPL through the Other Cleanup Activity (OCA) approach. OCA deferrals go to one of four types of entities (described below): states, federal agencies, tribes, or private parties.

- OCA deferral to a state places a site under that particular state's environmental regulations, as opposed to CERCLA authorities. In contrast to formal state deferrals, the OCA deferral to a state involves no formal EPA oversight other than periodic discussions between EPA regional officials and state officials. Since 2012, EPA guidance has indicated regions should have these discussions.
- OCA deferral to federal agencies places a site under that particular federal agency's oversight and authorities, according to EPA. Certain federal agencies, such as the Department of Defense, have responsibility and authority for some or all cleanups at their facilities.[18]

EPA assigns a status of "Other Cleanup Activity: Federal Facility Lead" to federal facilities that EPA tracks in its CERCLIS database and are being cleaned up outside of the NPL approach[19] (these sites are eligible for listing but are not listed on the NPL). EPA periodically checks in with other federal agencies on the status of cleanup work at these sites.

- OCA deferral to a tribe places the site under that tribe's environmental regulations. EPA periodically checks in with tribal regulators on the status of cleanup work at these sites.

- OCA deferral to private parties applies to certain sites where the cleanup is conducted by a private party.

EPA DEFERS OVERSIGHT OF A MAJORITY OF SITES ELIGIBLE FOR THE NPL TO APPROACHES OUTSIDE OF THE SUPERFUND PROGRAM

EPA most commonly addresses the cleanup of sites eligible for the NPL by "deferring" oversight to approaches outside of the Superfund program. EPA regions select the cleanup approach and defer oversight of more than half the sites eligible for the NPL to approaches outside of the Superfund program, primarily through OCA deferrals. Though OCA deferrals include the majority of NPL eligible sites, EPA's guidance on this approach is less detailed than guidance on other approaches.

EPA Regions Exercise Discretion in Selecting the Cleanup Approach for Sites

EPA provides regions with discretion in selecting the cleanup approach for a given hazardous waste site. According to the *Superfund Program Implementation Manual*—which lists EPA's Superfund program management priorities, procedures, and practices—each region is to select an appropriate cleanup approach after determining a site is eligible for the NPL. Officials in all 10 regions said that when they select cleanup approaches they attempt to use the most appropriate cleanup approach for a given site. For example, complex sites, such as contaminated waterways, may be more suited to the NPL approach than to deferral to a state cleanup program because EPA typically has more resources to oversee and manage such complex cleanups. Officials in four regions noted that states will sometimes request the NPL approach for large or complex sites.

EPA regions can establish their own processes for selecting a cleanup approach for a given site. Three of the 10 regions have some type of regional guidance related to their decision-making process. For example, Region 7 has guidance for its regional decision team that outlines the stakeholders within the region who will participate, when the team will meet, and how decisions are to be made at the meeting. Region 10 has guidance that focuses on how the region will prioritize sites that are eligible for the NPL. All of the regions may consult with relevant stakeholders across EPA programs about a given site, whether the regions have written guidance or not. These stakeholders might include staff from the office of regional counsel or the removal program. Five of the 10 regions use regional decision teams to evaluate sites that have been found eligible for the NPL and select which approach should be used to clean up the site. The other 5 regions do not use regional decision teams, opting instead for more informal decision-making processes or meetings on an as-needed basis. For example, in Region 5 there is a practice of coordinating between the region's long-term cleanup and removal programs on sites that may be of interest to both programs.

EPA officials said the regions consider many potentially relevant factors to select the appropriate approach for each site. The major factors influencing regional officials' choice of cleanup approach at a site include the preferences of the state regarding how the site will be addressed and the existence of a PRP that is willing to and capable of addressing the site. Specifically, officials in all 10 regions highlighted state preference as a factor they consider. State preference can be particularly important because EPA has a policy of obtaining state concurrence before listing a site on the NPL. According to Region 5 officials, if a state opposes an NPL listing, they will typically give preference to other approaches, such as an OCA deferral to the state or the SA approach. In addition to state preference, officials in 9 of 10 regions said that the existence of a willing and capable PRP can be a factor in determining the cleanup approach. For example, a willing and capable PRP is necessary for the SA approach, which requires the PRP to conduct the cleanup under an agreement with EPA. EPA officials in one region said that the existence of a PRP can also be important for cleanups under state cleanup programs because states can have very limited funding to conduct cleanups on their own. State environmental officials from four states we contacted confirmed that they had limited or, in some cases, no state funding to conduct their own long-term cleanups.

Regional officials also identified other factors that can sometimes influence what cleanup approach the region will select. For example, officials in Region 5 noted that if the contamination presents an immediate threat to health and safety, they may use the Superfund removal program, which is more suited to a quick response than long- term cleanup approaches. Depending on the circumstances at a site, the removal program may be sufficient to deal with all of the contamination, or the site may need to be referred to a long-term cleanup approach for further work. Regional officials can also consider other relevant legal authorities that could apply to a site, such as RCRA. When a site is eligible for cleanup under both RCRA and Superfund, EPA policy provides that the agency generally will defer the site to the RCRA program for cleanup.[20]

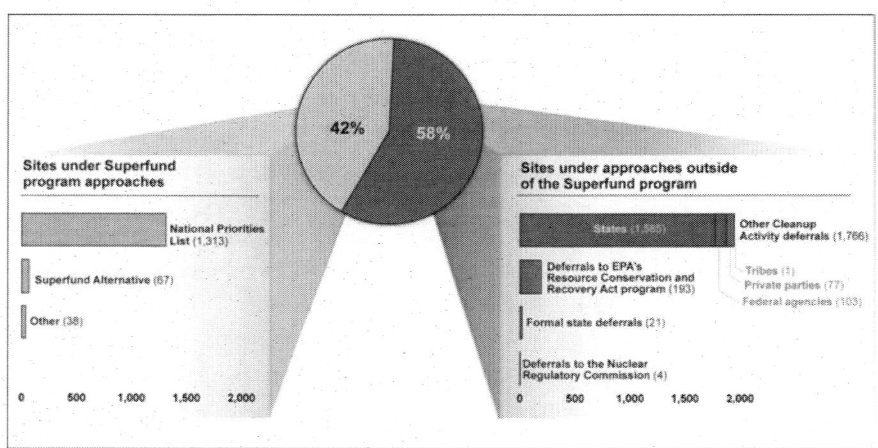

Source GAO analysis of EPA CERCLIS data.

Notes: Sites listed above only include sites that are currently active in CERCLIS and do not include sites that have been archived by EPA or sites where no further site response is required, such as sites deleted from the NPL.

Figure 4. Current Sites Identified as Eligible for the NPL Using Different Long-Term Cleanup Approaches, as of December 2012.

EPA Defers Oversight of More Than Half the Sites Eligible for the NPL to Approaches Outside of the Superfund Program

Among the 3,402 sites reported to the Superfund program in CERCLIS that EPA has identified as having contamination making them eligible for the NPL,[21] EPA deferred 1,984 sites to cleanup approaches outside of the Superfund program (see Figure 4). Sites under the Superfund program make up the 1,418 sites that remain, with the vast majority of those sites being addressed through the NPL.

EPA addresses more sites eligible for the NPL through the OCA deferral approach than any other cleanup approach: 1,766 of the 3,402 sites (52 percent). Moreover, because EPA deferred most of these 1,766 sites to states, OCA deferrals to states account for about 47 percent of all identified eligible sites.[22] EPA regions' use of OCA deferrals to states ranges widely, from 7 sites in each of three regions (6, 7, and 8) to 470 sites in Region 1 (see app. II for a breakdown of cleanup approaches by region). According to officials in Region 1, states in the region have mature environmental programs willing and capable of overseeing many sites, which makes the OCA deferral to states well suited to that region. In contrast, officials we spoke to in some regions noted that they needed to consider states' capacity to oversee a site before using the OCA deferral to states. Nine states have no OCA deferrals, and other states oversee hundreds of these sites, with the most in Massachusetts (247 sites), New Jersey (221), and California (180). Environmental officials in several states we contacted confirmed that states' use of and experience with OCA deferrals can differ substantially. One state official noted that these differences are likely related to how industrialized a state may be and the extent of cleanup programs in a given state. OCA deferrals to federal agencies, private parties, and American Indian tribes account for an additional 181 sites. OCA deferrals to federal agencies primarily involve military sites: 76 percent of these deferrals were to the Army, Navy, or Air Force. In addition, a majority of OCA deferrals to private parties come from Florida where, on the basis of a state law, PRPs can conduct cleanup without any formal agreement or order from the state, according to Region 4 officials. PRPs conducting such cleanups must submit regular reports to the state on their progress, and the state reserves the right to take the PRP to court, if necessary.

EPA currently addresses 1,418 sites (42 percent of those identified as eligible for listing on the NPL) through approaches under the Superfund program—most commonly, through listing the sites on the NPL. Specifically, sites listed on the NPL account for 1,313 sites, over 90 percent of sites under the Superfund program.[23] According to officials in one region, EPA has access to more resources than states and typically addresses sites that require greater or more specialized resources through the NPL approach. For example, regional officials noted, states face different limitations that can prevent them from pursuing cleanup under their programs including: technical capacity, legal resources, and financial resources. In addition, EPA officials in four regions noted examples where a state environmental program requested that the Superfund program pursue NPL listing because the state was having trouble getting a PRP to cooperate or the PRP went bankrupt.

In addition to listing sites on the NPL, EPA also oversees the long-term cleanup of sites through two other approaches under the Superfund program. First, the Superfund program currently oversees 67 sites under the SA approach.[24] Second, EPA oversees at least 38 other sites with long-term cleanups under the Superfund program for which EPA has no documented definition and no consistently applied method of counting. EPA officials provided different estimates of the number of such sites. One EPA official provided a method to identify these sites based on a code in CERCLIS, which resulted in the 38 sites listed above. However, another EPA official provided us a list of 35 such sites that had reached the remedial action phase. Of these 35 sites, 12 matched the 38 sites identified by the code in CERCLIS. In addition, 16 of the sites on the list of 35 had a code of "status undetermined" or had no code at all. EPA regional officials also identified other specific sites under the Superfund program, but some of those sites could not be identified by the code in CERCLIS, were not on the list of 35, or had no code at all. As of December 2012, 270 sites had the "status undetermined" code, and 101 had no such code in CERCLIS, making it impossible to determine the exact number of sites that EPA oversees under the Superfund program that are not being addressed under either the NPL or SA approaches. Tracking of these sites is discussed later in this report.

EPA addressed the remaining sites eligible for the NPL through different deferral approaches, primarily through deferrals to the RCRA program. Specifically, deferrals to the RCRA program account for 193 of these remaining sites (89 percent).[25] Aside from these deferrals to the RCRA program and a few deferrals to NRC, EPA deferred 21 sites to state programs using the formal state deferral approach in 4 of its 10 regions. EPA officials said that the OCA deferral to states approach has largely replaced the formal state deferral approach, and EPA does not anticipate using the formal deferral approach much in the future.

EPA Has Less Detailed Guidance for OCA Deferrals than for Other Less Commonly used Deferral Approaches

As discussed above, EPA addresses more sites through OCA deferrals than any other approach but has less guidance to define this approach or how deferral decisions should be documented than for its other deferral approaches. Unlike OCA deferrals, EPA has guidance or other documents outlining the process for deferrals to the RCRA program, deferrals to the NRC, and formal state deferrals. These documents clearly define or provide mechanisms to define the roles of the Superfund program and the entity that will conduct oversight at the site. For example, guidance for the formal state deferral approach specifies that the EPA region and the state should enter into a memorandum of agreement in which they clarify mutual expectations for their interaction and each party's responsibilities at deferred sites.[26] After the deferral, the region continues to review the state's progress and conduct any other activities required by its individual agreements with the state in each case.[27]

In contrast, EPA has not issued guidance focused on OCA deferrals that clearly defines the different types of OCA deferrals or what detail would be sufficient or appropriate to support its decisions at these sites. Instead, EPA describes OCA deferrals in the *Superfund Program Implementation Manual* (which is updated annually). EPA recently added to its instructions in the manual regarding sites with OCA deferrals. Specifically, in its 2012 version of the manual, EPA added more language explaining that there is to be no continuous and substantive involvement on EPA's part while cleanup work is ongoing at OCA deferral sites. In addition, in this version of the manual, EPA added an instruction for regions to check on the status of OCA sites periodically. Officials in EPA regions noted that they use different approaches for tracking OCA sites; for example, for an OCA deferral to states, EPA regions' tracking activities range from checking state websites to meeting with states to receive status updates every 3 months. Officials in some regions noted that they will need to modify their processes to meet this new instruction.

Even with EPA's additions to the manual, the available instruction does not clearly define each type of OCA deferral, particularly OCA deferrals to private parties, which has resulted in inconsistent identification of those OCA deferrals by different regions. While the manual defines OCA deferrals generally, it does not define each type of OCA deferral. When asked to define OCA deferrals to private parties, Superfund program officials in headquarters referred us to EPA regional officials for more information, and officials in 6 of 10 EPA regions were unsure about how to define OCA deferrals to private parties or how they should be used.[28] Moreover, officials in another 6 regions confirmed that some sites identified as OCA deferrals to private parties in CERCLIS should have been identified as OCA deferrals to states. Without clearer guidance defining the different OCA deferrals, EPA cannot be reasonably assured that it is consistently tracking its OCA deferral sites in CERCLIS, which can make it difficult to identify what entity is responsible for conducting oversight at the site.

In addition, the manual instructs regions to track OCA deferrals and completion of cleanups at these sites but does not clearly specify the documentation required to support these actions. The manual provides overarching program management priorities, procedures, and practices for the Superfund program. For OCA deferrals, the manual explains how EPA regions should identify the deferral date of an OCA site and the date of completion of cleanup at that site,[29] but it provides little detail on what type of documentation would be acceptable to support these determinations. For example, according to the manual, the deferral date of a site entered into CERCLIS is supposed to be "supported by existing documentation," described as "documentation between EPA and the non- EPA party leading the cleanup," with no further detail of what documentation would be appropriate or sufficient. In addition, the instruction for entering the date for completion of the cleanup refers to required documentation, without further clarification about what documentation is needed from the entities conducting oversight. Regional officials told us that, in practice, the amount and type of documentation regions collect to support OCA deferrals covers a broad range, including no written documentation, an e-mail from a state official, letters from state officials attesting to the cleanup, or a copy of the legal order or agreement between the state and PRP. Similarly, regions relied on different forms of documentation, including various e-mails, letters, or reports from state officials to document the completion of cleanup at OCA deferral sites. Officials in three regions reported that there was no consistent standard for documentation within their region. Moreover, Region 9 officials noted that the region had not tracked the completion of cleanups at OCA deferrals in CERCLIS in the past and may have no documentation for some of its older OCA deferral sites. Without guidance that details the documentation needed to support regions' OCA deferral decisions, EPA cannot be reasonably assured that its regions' documentation will be appropriate or sufficient to verify that these sites have been deferred or have completed cleanup. EPA officials noted they were working on additional guidance for OCA deferrals. However, these officials said that development of the guidance was in the planning stage; therefore, a draft of this guidance, detailed information on what will be included in the guidance, or a planned issuance date for the guidance, were not yet available.

EPA provides the least detailed guidance for the small number of sites that are undergoing long-term cleanup under the Superfund program outside of the SA and NPL approaches. Such sites do not have specific guidance at the program level, regional level, or a section in the *Superfund Program Implementation Manual* describing how they should be defined or tracked. In contrast, EPA has developed instructions in the manual for how to track sites cleaned up under the SA approach. EPA also has guidance for the NPL approach, such as how the agency should propose, list, and delete sites from the NPL. EPA officials noted that sites that are cleaned up under the other Superfund program approach often involve unique situations, making it difficult to establish any guidance that would cover all possible situations. For example, one of these sites is using a hybrid approach under both RCRA and CERCLA authorities, according to an EPA official. However, in 10 cases, regional officials described these sites as standard cleanups under CERCLA authority that used standard procedures. While there are unique and standard cases among sites being cleaned up under the other Superfund program approach (i.e., outside of the SA and NPL approaches), EPA officials could not provide a reliable estimate of these other sites because the agency has no consistently applied method for counting them. Without a method to identify and track such sites, EPA headquarters has no way to determine the extent to which regions use this approach or evaluate regions' use of this approach. As a result, it will be difficult for EPA headquarters to hold regions accountable for using the approach.

PROCESSES FOR IMPLEMENTING THE SA AND NPL APPROACHES DIFFER IN A FEW SIGNIFICANT WAYS

The processes for implementing the SA and NPL approaches have similarities, but also several differences, some of which EPA has accounted for through specific provisions in its agreements with PRPs at SA agreement sites. However, some sites may not benefit from EPA's efforts to account for these differences. Furthermore, the agency's tracking and reporting of SA agreement sites differs significantly from its tracking and reporting of NPL sites. Using the SA approach at sites has certain potential advantages for EPA and some PRPs and states, but communities' views on this approach are mixed.

The SA and NPL Processes are Similar in Many Ways and EPA Has Accounted for Some Differences between Them

The processes for implementing the SA and NPL approaches have many similarities. According to the agency's SA guidance, at its SA agreement sites, EPA is to generally act in accordance with the practices normally followed at sites listed on the NPL. For example, according to EPA guidance, SA agreement and NPL sites should follow the same investigation and cleanup processes, including the phases and milestones of long-term cleanups shown earlier in Figure 3. EPA regions should also use the same response techniques, standards, and guidance for SA agreement sites as they do for NPL sites. According to EPA's guidance, SA agreements should eventually achieve cleanup levels that are comparable to those required at NPL sites. EPA regions should also take steps to ensure equivalency between the SA and NPL approaches in the absence of NPL listing.

Despite these similarities, there are certain differences in the overall processes and EPA's authority under the NPL and SA approaches. Through specific provisions in its SA agreements with PRPs, EPA has sought to make the two approaches comparable by accounting for the following four key differences:

- First, EPA has the authority to pay for remedial actions only at sites listed on the NPL.[30] To account for this difference, SA agreements include a provision to help ensure cleanups are not delayed by a loss of funding if the PRP ceases work during the remedial action phase of cleanup. Specifically, this provision requires the PRP to obtain a readily available source of funds that the agency can use if it needs to take over the cleanup work. EPA can use those funds to continue the work while the agency lists the site on the NPL, if necessary.

- Second, EPA is authorized to provide technical assistance grants that help communities participate in decision making only at sites that are listed or proposed for listing on the NPL. An initial grant of up to $50,000 is available to qualified community groups so they can contract with independent advisors to help the community understand technical information about the site. EPA includes a provision in SA agreements to help ensure that a community's opportunity to receive technical assistance at an SA agreement site is comparable to that at an NPL site. This provision requires the PRPs, with EPA oversight, to administer and fund a technical assistance plan, under which a qualified community group can receive up to $50,000 for the same purposes as a technical assistance grant from EPA.

- Third, if a PRP were to clean up an SA agreement site to the extent that it no longer scored at least 28.50 on the Hazard Ranking System, according to EPA, it might lose the option of listing the site on the NPL, a concern that is not present when a site is listed on the NPL. To prevent this, SA agreements state that the PRP will not challenge listing the site on the NPL if a partial cleanup of the site results in changed site conditions. EPA officials noted that this provision gives the agency assurance that it can step in and clean up the site under the NPL approach if the PRP were to default on the SA agreement.

- Fourth, CERCLA states that an action for natural resource damages (NRD) at NPL sites must start within 3 years after completion of the remedial action.[31] This period is longer than the general statute of limitations for NRD claims, which states that an action must start within 3 years after the discovery of the loss and its connection with the contamination. SA agreements contain a provision that clarifies that the longer statute of limitations for NPL sites also applies to SA agreement sites.

Even with EPA's efforts to achieve equivalence of SA agreement and NPL sites through these provisions, some sites may not benefit from these efforts because EPA regions have entered into agreements with PRPs at sites that they said were likely eligible for the SA approach without following the SA guidance. Agreements at such sites may not, for example, ensure that a community has access to funds to pay for technical assistance or that remedial action can continue if a PRP stops cooperating. Officials from some EPA regions told us they have continued to enter into agreements with PRPs since 2002 without following the SA guidance. We identified six sites where this has occurred as follows:

- In Region 7, officials entered into an agreement with a PRP to conduct remedial design and remedial action at a site. Regional officials stated that the SA approach, which can be suggested for a site by the PRP or the region, never came up during their discussions with the PRP.

- In Region 10, officials stated that the agreements they had entered into with PRPs at five sites might qualify for the SA approach but that, at the time they entered into the agreements, the officials had not focused on whether the agreements met the SA criteria; rather, they were focused on obtaining enforceable agreements.

According to EPA headquarters officials, if regions are going to conduct a long-term cleanup under the Superfund program at a site, but not list it on the NPL, the agency prefers regions to use the SA approach. EPA headquarters officials said that they believed this preference was implicit in the agency's SA guidance and stated they discussed this preference with regional officials at periodic meetings; however, they also acknowledged that this preference is not stated explicitly anywhere in guidance for the regions. If regions continue to enter into agreements for some sites without following the SA guidance, these sites may be denied some of the advantages built into the SA agreements to ensure that the cleanups will be comparable to those under the NPL approach.

EPA's Tracking and Reporting of Certain Aspects of the Process under the SA Approach Differ Significantly from That under the NPL Approach

Some differences remain between the way EPA tracks sites under the SA and NPL approaches. In CERCLIS, EPA tracks sites' status in relation to the NPL regardless of any changes in cleanup approach. Specifically, sites that have been proposed for listing on the NPL, are currently on the NPL, have been deleted from the NPL, or have been removed from proposal can always be identified as such in CERCLIS, which allows EPA to accurately identify sites that are or have been on the NPL. In contrast, EPA cannot similarly track an SA agreement site as such if it is subsequently listed on the NPL.[32] Specifically, EPA currently tracks SA agreement sites through a single database code identifying only that a site has an SA agreement, and the identifying code is not maintained in the database if the site is later added to the NPL. The agency has not clarified in its guidance when to leave this SA identifying code in place, and when to remove it, even though the EPA IG recommended in a 2007 report that EPA develop specific instructions on when to use the SA designation and update the *Superfund Program Implementation Manual* (which is updated annually) to incorporate these instructions.[33] According to the IG report, these instructions should specify that the SA code should not be removed even if the site is cleaned up or proposed for the NPL, so that controls over documentation of sites with SA agreements can be maintained. As the EPA IG pointed out, absence of guidance can result in poor quality data on the SA universe. While EPA indicated in 2010 that it would implement this recommendation,[34] the 2012 manual does not include any instructions about maintaining the SA code. Because EPA has not implemented the IG's recommendation, the manner in which the agency tracks the identity of SA agreement sites in CERCLIS is incomplete. For example, while an EPA website identifies all sites that have or have had SA agreements, three sites that had SA agreements and were later added to the NPL cannot be identified in CERCLIS as having had SA agreements. As a result, all sites that have had SA agreements are not identifiable in CERCLIS, which may hamper EPA's ability to effectively manage long-term cleanups and track outcomes at SA agreement sites.

Furthermore, the standards for specifying what documentation is sufficient to support the Hazard Ranking System score of SA agreement sites are less clear than those for NPL sites. When sites are proposed for listing on the NPL, EPA procedure requires they have a Hazard Ranking System documentation record—a specific document that includes detailed justification for the Hazard Ranking System score. In contrast, both the 2004 and 2012 SA guidance state that EPA should have "adequate documentation" supporting a Hazard Ranking System score of 28.50 or higher but do not define what is meant by "adequate" documentation or provide criteria for assessing adequacy. The guidance documents specify that regions may rely on a draft Hazard Ranking System documentation record or "other adequate documentation," but do not provide an explanation of what other documentation might be adequate. EPA headquarters officials told us that documentation of a preliminary calculation of the Hazard Ranking System score during the initial assessment phases would qualify as adequate, and said that this has been discussed with regional officials during periodic meetings. EPA officials acknowledged, however, that this interpretation of the guidance has not been included in any written guidance to the regions. As the EPA IG pointed out in its 2007 report, consistent and reliable documentation of Hazard Ranking System scores at SA agreement sites is an internal control to ensure compliance with the SA guidance and approach.[35] Under the federal standards of internal control, agencies are to clearly put in writing (i.e., in management directives, administrative policies, or operating manuals) internal controls, such as this interpretation of the guidance, and have them readily available for examination.[36] Without more specific written guidance, EPA regional officials may not develop adequate documentation of Hazard Ranking System scores at SA agreement sites.

In addition to the differences in its tracking, EPA has not reported the agency's performance on the progress of cleanup at SA agreement sites as it has for NPL sites. EPA reports achievements at NPL sites, including completion of some phases of the cleanup process, as part of the agency's implementation of provisions under GPRA, which generally aims to hold federal agencies accountable for using resources wisely and achieving program results. Two of the Superfund program's three GPRA performance measures—sites where human exposure is under control and sites that are ready for their anticipated use—refer only to NPL sites. One additional performance measure tracks the completion of the initial assessment phases, which generally precede EPA's decision about which cleanup approach to use at a site, including the SA or NPL approach. EPA's Office of Solid Waste and Emergency Response, which manages the Superfund program, reports these performance measures for NPL sites in several annual reports available on EPA's website.[37] However, EPA does not include in these reports the cleanup milestones reached at SA agreement sites, such as how many SA sites have human exposure under control. The EPA IG recommended in 2007 that EPA track and report the same GPRA performance measures at SA agreement sites as it does at NPL sites.[38] As the IG reported, by measuring and tracking all performance measures at SA agreement sites, EPA could demonstrate the outcomes of the Superfund program's work and provide an incentive to regions by more thoroughly accounting for their performance. In 2010, EPA indicated that it would implement the IG's 2007 recommendation to track and report all Superfund GPRA performance measures at SA agreement sites using an annual report.[39] EPA officials noted that the agency has begun tracking Superfund performance measures for SA agreement sites, but they acknowledged that EPA is not reporting these results publicly. Until the agency reports performance information on the progress of cleanup at SA agreement sites as it does for NPL sites, EPA is not providing the public and Congress with a full picture of SA agreement sites. Without such information, Congress lacks complete information on the progress of the Superfund program to inform its legislative actions, including appropriations.

The SA Approach Has Potential Advantages for EPA, Some PRPs, and States, but Communities' Views on its Benefits Are Mixed

Using the SA or the NPL approach can have advantages or disadvantages for the parties involved, including EPA, PRPs, states, and communities. Specifically, using the SA approach generally allows EPA to avoid at least some of the cost and time associated with listing a site on the NPL. For example, NPL listing requires preparation of a Hazard Ranking System documentation record, which is not required for sites with SA agreements.[40] EPA officials estimated each such record costs an average of about $65,000 to prepare. In addition, when EPA decides to propose a site for listing on the NPL, the agency sometimes conducts an expanded site inspection if further information is necessary to document a Hazard Ranking System score. EPA officials estimated this step costs about $92,000 on average. In addition, to list a site on the NPL, EPA has to work through the formal listing process, including issuing notices in the *Federal Register* with a public comment period. This process takes time to complete, which may affect the progress at the site. In Region 3, EPA officials stated that the volume of comments received on a particular site proposed for the NPL, in addition to the likelihood of litigation from one or more parties if the site were finalized on the NPL, led the region to address the site through the SA approach.

Some EPA regions have seen the advantages of using the SA approach more than others. As shown in Figure 5, of the 67 SA agreement sites, 57 sites, or 85 percent, are in EPA Regions 4 and 5.

Differences in usage of the SA approach among regions relate to a region's specific circumstances and preferences. According to EPA headquarters officials, Regions 4 and 5 had early experience with SA agreements and may have been more comfortable in starting new ones as a result. These two regions have also listed many sites to the NPL since the SA approach was formalized in 2002. Region 4 officials told us that they have found that the SA approach is best suited to sites with one or two PRPs and no questions about the PRPs' liability or ability to conduct the investigation or cleanup. They said they have also found that it is helpful when a PRP has a financial interest in finishing the cleanup quickly, as in the case of a potential redevelopment project at a site. Other regions have used the SA approach in limited circumstances. For example, officials in Region 9 described one case in which they pursued the SA approach because the state did not want a particular site listed on the NPL. Two regions—Regions 1 and 2—have never used the SA approach. Officials in Region 1 explained that few sites that have willing and capable PRPs and are eligible for the NPL come to the region's attention because state programs prefer to take on oversight of such sites. Region 2 officials said they did not see a reason to use the SA approach—if a site's contamination is severe enough, the region will propose the site to the NPL, unless the state is addressing the site.

Using the SA approach allows PRPs to avoid the perceived stigma associated with an NPL site, according to EPA officials. Sites with SA agreements have to meet all of the qualifications of NPL sites, and thus may have contamination that is just as severe, but the potential stigma of NPL listing appears to influence PRPs. Officials in 7 of 10 regions mentioned the stigma of an NPL site as a concern for PRPs. Concerns about this stigma may also arise when a company is to be sold and does not want to list an NPL site as part of its liabilities, according to EPA regional officials. Related to this stigma, EPA officials said they believed that avoiding listing on the NPL may help local government officials and PRPs in some cases, such as facilitating a site's redevelopment or its financing. Previous reports have also pointed to the potential stigma of NPL listing as motivation for pursuing a different cleanup approach.[41] For example, an assessment of the effectiveness of the SA approach in Region 4 (hereafter referred to as the Region 4 study) found that sites using the SA approach may have a higher potential for redevelopment than comparable NPL sites if avoiding this stigma increases PRPs' financing options and their willingness to redevelop.[42]

In addition, some states generally prefer that EPA not list sites on the NPL, according to EPA officials, which makes the SA approach more appealing. According to EPA policy, EPA typically obtains a state's concurrence before listing a site on the NPL. Officials in all 10 EPA regions mentioned the states' views as one of the factors they used to determine whether to pursue an NPL listing or other approaches. Moreover, officials in 4 of 10 regions said there were states in their region that were generally reluctant to have EPA list sites on the NPL. For example, Region 9 officials said two of their states generally do not want EPA to list sites on the NPL; specifically, one of these states wanted to avoid the associated stigma of having NPL sites in the state.

The SA approach also has advantages and disadvantages for communities. According to an EPA official, it may be easier for communities to obtain technical assistance funds from PRPs at SA agreement sites than to obtain the equivalent funds from EPA at NPL sites. This official said obtaining funds from PRPs at SA agreement sites often involved the absence of a "match" requirement as well as fewer paperwork requirements for the communities because the technical assistance plans do not have to follow federal grant requirements. However, under the SA approach, communities have no opportunity for a formal comment process on EPA's selection of the SA approach itself, as they do under the NPL approach. Specifically, when EPA proposes a site for the NPL in the *Federal Register*, the public has 60 days to comment on the proposed listing. EPA then responds in writing to significant public comments in conjunction with the final *Federal Register* listing announcement of the site. No such opportunity exists when EPA decides to enter into an SA agreement at a site, although EPA provides numerous opportunities under the SA approach for communities to comment on the cleanup process.

Communities also may have mixed reactions to the SA approach for other reasons as well. According to EPA officials, communities may have concerns about the SA approach and may require outreach from the agency to explain the approach. For example, at one site in Region 5, the region expanded its outreach efforts after some community members protested the use of the SA approach at the site. A regional official explained that some individuals in the community believed the site would not follow the same cleanup process as an NPL site. Some community members may support listing on the NPL over the SA approach to bring increased attention to a site, helping to ensure its cleanup. Other regional officials said other community members may be more open to the SA approach and oppose listing on the NPL for fear of its effect on property values. The Region 4 study confirmed that the SA approach is often considered advantageous by community members and leaders concerned about property values and stigma.[43] However, this report also found that other community members require confirmation that the process will not result in more limited resources or reduced remediation compared to listing on the NPL.

SA AGREEMENT SITES SHOWED MIXED RESULTS IN COMPLETING THE CLEANUP PROCESS WHEN COMPARED WITH SIMILAR NPL SITES

For sites with agreements from June 2002 through December 2012, SA agreement sites and similar NPL sites we selected showed mixed results in the time needed to complete negotiations for agreements, specific cleanup activities, and achieving the construction completion milestone (see app. I for more details on our objectives, scope, and methodology and app. III for more information on our results). Specifically, SA agreement and NPL sites in our analysis showed mixed results in the average time to complete negotiations with PRPs and for specific cleanup activities, such as remedial investigation and feasibility studies, remedial designs, and remedial actions. In addition, a lower proportion of SA agreement sites have reached construction completion compared with similar NPL sites. SA agreement sites tend to be in earlier phases of the cleanup process because the SA approach began more recently than the NPL approach.

Sources: GAO analysis of EPA data; Map Resources (map).

Figure 5. Number of SA Agreement Sites by EPA Region, as of December 2012.

For agreements finalized from June 2002 through December 2012 at sites in our analysis, SA agreement and similar NPL sites showed mixed results in the length of time to complete negotiations, with SA agreement sites taking about as long as similar NPL sites for remedial investigation and feasibility study negotiations and less time for remedial design and remedial action negotiations. EPA regional officials confirmed that negotiations can be faster at SA agreement sites because the PRPs are more cooperative. For example, Region 4 officials highlighted one SA agreement site where the PRP pushed for a quicker negotiation process by turning in documents ahead of deadlines, unlike many other PRPs. In another case, Region 5 officials said they negotiated three SA agreements for remedial investigations and feasibility studies covering 19 sites of a similar nature with the same PRP. Region 5 officials noted that these negotiations were particularly smooth and cooperative. Moreover, the Region 4 study also found, based on interviews with PRPs and EPA officials, that the tone of SA negotiations is more productive than at NPL sites.[44] However, given the relatively limited number of negotiations for both NPL and SA agreement sites in our analysis, the differences in the average length of negotiations cannot be attributed entirely to the type of approach used at each site.

The SA agreement and similar NPL sites in our analysis showed mixed results in the length of time it took to complete specific cleanup activities,[45] with SA agreement sites taking substantially longer for remedial investigations and feasibility studies on average and about the same time for remedial designs and remedial actions. While SA agreement sites took substantially longer on average than NPL sites to complete remedial investigations and feasibility studies, these differences do not appear to be exclusively attributable to the SA and NPL approaches. For example, several remedial investigations and feasibility studies at SA agreement sites took a long time to complete due to individual circumstances at the site, such as dealing with a proposal to sell on-site materials to a manufacturing company, late participation from PRPs in the process, or coordination with other cleanup efforts. SA agreement sites and NPL sites in our analysis took about the same time on average to complete remedial designs and remedial actions.

A lower proportion of SA agreement sites have reached construction completion compared with similar NPL sites in our analysis (see Figure 6).

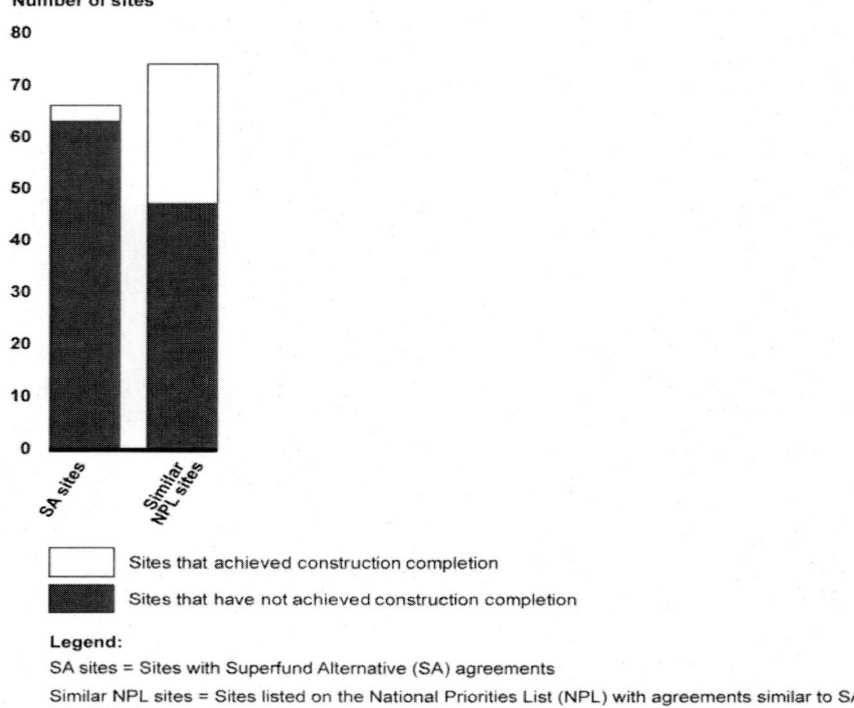

Source GAO analysis of EPA data.

Figure 6. Construction Completions at SA Sites and Similar NPL Sites.

We selected NPL sites for our analysis that had agreements put in place between EPA and PRPs from June 2002 through December 2012, as SA agreement sites do. According to EPA officials, however, because the SA approach began in 2002, and the NPL approach was initially authorized in 1980, more SA agreement sites began cleanup later than NPL sites and, therefore, are in earlier phases of the cleanup process. For example, 23 percent of the sites in our NPL comparison group have ongoing remedial investigations and feasibility studies—a phase that precedes selection of the remedy and remedial action—while almost 60 percent of SA agreement sites are in this phase. Since EPA began implementing the NPL approach over 30 years ago, there were more sites ready to negotiate agreements during the period of our analysis for later phases in the cleanup process, which lead to construction completions.

CONCLUSION

EPA regional officials are responsible for choosing the appropriate long-term cleanup approach for sites with contamination that makes them eligible for the NPL. To do so, they select from among several approaches, including deferring responsibility for the oversight of site cleanup outside of the Superfund program. Of these sites deferred outside of the Superfund program, EPA has deferred about 1,800 sites through OCA deferrals—more sites than any other approach—but the agency has not issued guidance focused on this long-term cleanup approach. Instead, EPA describes OCA deferrals in the *Superfund Program Implementation Manual,* which does not clearly define each type of OCA deferral, particularly OCA deferrals to private parties. This has led to inconsistent coding of OCA deferrals in CERCLIS by different regions. Moreover, EPA's guidance does not specify in detail the documentation regions should have to support their decisions on OCA deferrals or completion of cleanup at these sites. As a result, EPA regions collected varying types and amounts of documentation—including, in some cases, no documentation—to support OCA deferrals. EPA officials noted they were currently working on additional guidance for OCA deferrals, but they had not set an issuance date for this guidance. Without clearer guidance on OCA deferrals, EPA does not have reasonable assurance that it can consistently track its OCA deferral sites in CERCLIS or that its regions' documentation will be appropriate or sufficient to verify that these sites have been deferred or have completed cleanup. In addition, EPA officials could not provide a reliable estimate regarding the number of sites with long-term cleanups under the Superfund program that are being cleaned up through approaches other than the NPL and SA approaches—the "other" Superfund program sites—because there is no consistently applied method for tracking them. While the agency's estimates of the number of such sites is relatively small, without a method to identify and track such sites, it is difficult for EPA headquarters to determine the extent to which regions use this other approach under the Superfund program, evaluate regions' use of this approach, or hold regions accountable for using this approach.

Furthermore, EPA guidance has made clear since 2002 that the agency should try to make SA agreement sites equivalent to NPL sites in terms of the level of cleanup achieved, among other things. EPA has largely accomplished this through adherence to the Superfund cleanup process and by adding certain provisions to SA agreements to address key differences between the NPL and SA approaches. The agency has not clarified to regions in its guidance that the

SA approach is the preferred approach for long-term cleanup of sites under the Superfund program not listed on the NPL. Without clear guidance, agreements at such sites may be denied some of the advantages built into the SA agreements to ensure that the cleanups will be comparable to those under the NPL approach. Also, while EPA accurately identifies NPL sites in CERCLIS, the agency cannot do the same for SA agreement sites because it has not clarified in writing when the database code that identifies sites with SA agreements should remain in place and when it should be removed. In addition, EPA's standards for specifying what documentation is sufficient to support the Hazard Ranking System score at SA agreement sites are less clear than those for NPL sites. Unless EPA improves its tracking of SA agreement sites and clarifies its policies, its ability to effectively track outcomes of the SA approach at these sites and manage long-term cleanups at sites under the Superfund program may be hampered. Finally, while EPA reports performance information for NPL sites under GPRA, it does not report performance information on the progress of cleanup at SA agreement sites in an equivalent manner. Without such information on SA agreement sites, Congress lacks complete information on the progress of the Superfund program to inform its legislative actions, including appropriations.

RECOMMENDATIONS FOR EXECUTIVE ACTION

To improve the Superfund program's management of sites with contamination that makes them eligible for the NPL, including management of the SA approach and deferrals of cleanup oversight to other entities, we recommend that the Administrator of EPA take the following four actions:

- Provide guidance to EPA regions that defines each type of OCA deferral and what constitutes adequate documentation for OCA deferral and completion of cleanup.
- Develop a method for EPA headquarters to identify and track other sites with long-term cleanups under the Superfund program (i.e., those that are outside of the NPL and SA approaches).
- Update EPA's written policies on SA agreement sites, including taking steps such as clarifying whether the SA approach is EPA's preferred approach for long-term cleanup of sites under the Superfund program and outside of the NPL, specifying what documentation is sufficient to support the Hazard Ranking System score at SA

agreement sites, and defining when the database code that identifies sites with SA agreements should remain in place.

- Report performance information on the progress of cleanup at SA agreement sites in a manner that is equivalent to such reporting for NPL sites.

AGENCY COMMENTS AND OUR EVALUATION

We provided a draft of this report to EPA for review and comment. In written comments, EPA agreed with the report's recommendations and stated that it believes the report contains substantial useful information. Regarding the first recommendation, EPA stated that it added more detail on OCA tracking in its fiscal year 2012 *Superfund Program Implementation Manual,* but it acknowledged that more guidance is needed. Regarding the second recommendation, EPA stated that it agreed with the recommendation without further comment. Regarding the third recommendation, EPA said that it will clarify that the SA approach is generally the agency's preferred enforcement approach for CERCLA non-NPL sites that are "NPL-caliber," where feasible and appropriate. Finally, regarding the fourth recommendation, EPA stated that it agrees with this recommendation as it pertains to reporting under GPRA and provided further information on how EPA reports measures at SA agreement sites. EPA also provided technical comments on the draft report, which we incorporated as appropriate.

David C. Trimble
Director, Natural Resources and Environment

APPENDIX I: OBJECTIVES, SCOPE, AND METHODOLOGY

This appendix provides information on the scope of the work and the methodology used to examine (1) how the Environmental Protection Agency (EPA) addresses the cleanup of sites it has identified as eligible for the National Priorities List (NPL), (2) how the processes for implementing the Superfund Alternative (SA) and NPL approaches compare, and (3) how SA agreement sites compare with similar NPL sites in completing the cleanup process.

To examine how EPA addresses the cleanup of hazardous waste sites with a level of contamination that makes them eligible for the NPL, we analyzed applicable federal statutes and EPA regulations and guidance to determine the available approaches to address sites that are reported to the Superfund program. We then obtained and analyzed data from EPA's Comprehensive Environmental Response, Compensation, and Liability Information System (CERCLIS), the Superfund program's database, as of December 2012. Specifically, we analyzed EPA's CERCLIS database to determine how many sites EPA currently classified as undergoing long- term cleanup under each approach, both nationally and by each of EPA's 10 regions.[46] According to EPA officials, sites under each of the long-term cleanup approaches would have a Hazard Ranking System score of at least 28.50, otherwise the site would have been classified as "no further remedial action planned." Thus, all sites identified as being under a long- term cleanup approach were considered to have contamination making them eligible for the NPL. This analysis involved the review of EPA's non- NPL status code, NPL status code, and SA code. In addition, we conducted semistructured interviews with officials in all 10 EPA regions to understand each region's processes for selecting among long-term cleanup approaches and why regions used the various approaches. We also obtained relevant supporting documentation from these regional officials. In addition, we interviewed EPA headquarters officials about the assessment process and cleanup approaches. Finally, we interviewed a nonprobability, convenience sample of officials from 13 state cleanup programs who were familiar with available cleanup approaches. The convenience sample consisted of representatives from state environmental departments taking part in an Association of State and Territorial Solid Waste Management Officials conference call who agreed to speak with us. Because this was a nonprobability sample, the results of our analysis cannot be generalized to all states; however, these officials provided important information about the cleanup process.

To compare the processes for implementing the SA and NPL approaches, including the cleanup process and EPA's oversight, we analyzed available documentation on the two approaches, including guidance and prior reviews. These reviews included an EPA Inspector General (IG) report on the SA approach, as well as several reports on the approach by EPA. We reviewed key findings and recommendations from the IG's report, as well as the evidence provided by EPA to demonstrate its implementation of the report's recommendations. We found the evidence to be sufficient to assess whether EPA had implemented these recommendations. In addition, we interviewed officials in all 10 EPA regions to determine how each region implemented the SA approach and obtained relevant supporting documentation. Finally, we interviewed EPA headquarters officials knowledgeable about the SA approach.

To compare how SA agreement sites and similar NPL sites complete the cleanup process, we identified SA agreement sites and constructed a comparison group of 74 NPL sites with agreements between EPA and potentially responsible parties (PRP) similar to those at SA agreement sites as follows:

- We identified 67 SA agreement sites using the SA code and added to that 3 SA agreement sites with their SA code removed after the site was listed on the NPL for a total of 70 SA agreement sites; we identified these three sites through our interviews with EPA officials. We then obtained data on the legal actions taken at these sites from EPA officials in the Office of Site Remediation Enforcement, which included all agreements at these sites. Based on discussions with EPA officials and the SA guidance, we isolated agreements at SA agreement sites by selecting: (1) agreements entered into between June 2002 (the date of the issuance of the first SA guidance) and December 2012; (2) administrative orders on consent or consent decrees; and (3) agreements involving a PRP-led combined remedial investigation and feasibility study, remedial design, or remedial action. After excluding four sites with SA codes that had SA agreements that were not relevant to our study, we had 66 SA agreement sites for our analysis.[47]
- We constructed our comparison group of 74 NPL sites starting with the approximately 1,300 sites on the NPL. Specifically, we identified the 702 sites with (1) a combined remedial investigation and feasibility study, (2) remedial design, or (3) remedial action led by a PRP. We requested data on the legal actions taken at these sites from EPA officials and identified agreements similar to SA agreements based on the date the agreement was entered into, the type of agreement, and whether it included PRP-led long-term cleanup actions. In addition, we dropped any NPL sites from Regions 1 and 2 from the analysis because neither region has used the SA approach.

To more precisely align the NPL comparison group with SA agreement sites, we analyzed, for SA agreements, the number of PRPs involved and estimated costs for PRP-led actions. According to EPA officials, SA agreement sites generally tend to have fewer PRPs. Based on this analysis and EPA's comments, we established thresholds for different variables that agreements in our NPL comparison group could not exceed. Specifically, NPL agreements could have: (1) no more than seven PRPs involved and (2) administrative orders on consent with estimated values between $100,000 and $5,000,000 or consent decrees with estimated values between $125,000 and $30,000,000.[48] These ranges covered the vast majority of SA agreements.

After we identified the NPL sites with agreements similar to SA agreement sites, we merged the data on the legal actions with cleanup action data for NPL and SA agreement sites. We kept (1) combined remedial investigation and feasibility studies, (2) remedial designs, and (3) remedial actions at sites if the action was explicitly listed as a remedy in an SA agreement or an SA-similar agreement (for NPL sites). We identified negotiations related to cleanup actions of interest by comparing the completion date of the negotiation with the completion date of the agreement in EPA's legal action data. For remedial investigation and feasibility study negotiations, we kept any negotiation with a completion date up to 180 days before the date of an administrative order on consent for that site. For remedial design and remedial action negotiations, we kept any negotiation with a completion date up to 2 years before the date of a consent decree.[49] After keeping these cleanup actions of interest, we computed the durations of specific cleanup activities by calculating the difference in months between the start and completion dates of identified actions included in CERCLIS. We then calculated the mean and median durations for the SA and NPL groups, as well as related ranges. We compared the means and medians of the durations to assess whether reported results are affected by a possible skewed distribution. We decided to report the median because it is less sensitive to extreme values and provides a better estimate of the "average" duration for this analysis. Because only three SA agreement sites had reached the construction completion milestone, we were unable to compare the groups across the entire cleanup process; instead, we compared completion of specific activities, such as remedial designs. The results of our analysis cannot be generalized to all NPL sites because the 74 sites were a subset of all NPL sites selected to be as similar as possible to SA agreement sites based on key characteristics related to cleanup durations such as having a PRP that agreed to conduct at least some part of the cleanup. The comparison group was created for purposes of assessing whether alternative approaches for addressing the long-term cleanup of hazardous waste sites under the Superfund program can make a difference in cleanup durations and not for making generalizations about the larger universe of all NPL sites.

We conducted additional analyses on our SA and NPL groups to determine if there were any unaccounted distributional differences within each group that would materially affect our results. Specifically, we examined the sensitivity of our results to differences in regional distribution because the SA approach has different regional usage patterns than the NPL approach. While 85 percent of SA agreement sites are in Regions 4 and 5, only 34 percent of the similar NPL sites are in Regions 4 and 5. In one analysis, we restricted SA agreement and similar NPL sites to Regions 4 and 5, and the results were generally similar to the analysis using the full set of SA agreement and similar NPL sites.[50] In addition, we examined the sensitivity of our results to differences in the complexity of SA agreement sites and similar NPL sites measured through the distribution of megasites[51] and single operable unit sites in each group. The results for length of negotiations were not sensitive to differences between SA agreement sites and similar NPL sites in the distribution of megasites, though the results for the length of cleanup activities were somewhat sensitive to distributional differences.[52] The results for length of negotiation and cleanup durations were, in general, not sensitive to differences in the distribution of sites with one or more operable units.

To assess the reliability of the data from EPA's CERCLIS database used in this report, we analyzed related documentation, examined the data for errors or inconsistencies, and interviewed agency officials about any known data problems and to learn more about their procedures for maintaining the data. Where there were discrepancies in the data, we worked with EPA officials to clarify. For example, we identified certain SA agreement sites that did not appear to have agreements with long-term cleanup actions and reviewed these with EPA officials. Miscoded data were corrected, and EPA officials provided explanations for unique circumstances with certain agreements. We determined the data to be sufficiently reliable for calculating durations for completing different cleanup activities, including negotiations, at SA and NPL sites.

We conducted this performance audit from November 2011 to April 2013 in accordance with generally accepted government auditing standards. Those standards require that we plan and perform the audit to obtain sufficient, appropriate evidence to provide a reasonable basis for our findings and conclusions based on our audit objectives. We believe that the evidence obtained provides a reasonable basis for our findings and conclusions based on our audit objectives.

APPENDIX II: LONG-TERM CLEANUP APPROACHES BY EPA REGION

Tables 1 and 2 provide a breakdown of cleanup approaches by region. Table 1 shows the number of sites within each region that are being cleaned up under the various cleanup approaches. Table 2 shows each region's percentage of the total number of sites cleaned up under each approach

Table 1. Number of Sites by Region and Long-Term Cleanup Approach, as of December 2012

Region	NPL	SA	Other sites under Superfund program	OCA deferral to states	OCA deferral to federal agencies	OCA deferral to private parties	OCA deferral to tribes	Deferral to EPA's RC RA program	Formal state De ferrals	Deferral to the NRC	Total
1	100	0	0	470	4	1	0	2	0	0	**577**
2	215	0	8	247	5	0	0	5	0	0	**480**
3	170	2	1	119	22	5	0	23	0	2	**344**
4	185	23	11	289	7	55	0	71	5	0	**646**
5	244	34	3	111	22	5	0	13	0	0	**432**
6	90	1	0	7	0	3	0	20	0	0	**121**
7	68	1	2	7	0	0	0	7	8	1	**94**
8	53	1	1	7	1	6	0	11	1	1	**82**
9	113	3	1	222	37	1	1	40	0	0	**418**
10	75	2	11	106	5	1	0	1	7	0	**208**
Total	**1,313**	**67**	**38**	**1,585**	**103**	**77**	**1**	**193**	**21**	**4**	**3,402**

Source: GAO analysis of EPA data.

Notes: Sites listed above only include sites that are currently active in CERCLIS and do not include sites that have been archived by EPA or sites where no further site response is required, such as sites deleted from the NPL.

Table 2. Percentage of Sites by Region and Long-Term Cleanup Approach, as of December 2012

Region	NPL	SA	Other sites under Superfund program	OCA deferral to states	OCA deferral to federal agencies	OCA deferral to private parties	OCA deferral to tribes	Deferral to EPA's RCRA program	Formal state deferrals	Deferral to the NRC	Total
1	8%	0%	0%	30%	4%	1%	0%	1%	0%	0%	**17%**
2	16%	0%	21%	16%	5%	0%	0%	3%	0%	0%	**14%**
3	13%	3%	3%	8%	21%	6%	0%	12%	0%	50%	**10%**
4	14%	34%	29%	18%	7%	71%	0%	37%	24%	0%	**19%**
5	19%	51%	8%	7%	21%	6%	0%	7%	0%	0%	**13%**
6	7%	1%	0%	0%	0%	4%	0%	10%	0%	0%	**4%**
7	5%	1%	5%	0%	0%	0%	0%	4%	38%	25%	**3%**
8	4%	1%	3%	0%	1%	8%	0%	6%	5%	25%	**2%**
9	9%	4%	3%	14%	36%	1%	100%	21%	0%	0%	**12%**
10	6%	3%	29%	7%	5%	1%	0%	1%	33%	0%	**6%**
Total	**100%**	**100%**	**100%**	**100%**	**100%**	**100%**	**100%**	**100%**	**100%**	**100%**	**100%**

Source: GAO analysis of EPA data.

Notes: Percentages may not add up to 100 because of rounding. Sites listed above only include sites that are currently active in CERCLIS and do not include sites that have been archived by EPA or sites where no further site response is required, such as sites deleted from the NPL.

APPENDIX III: DATA ANALYSIS OF SA AGREEMENT SITES AND SIMILAR NPL SITES

In this appendix, we discuss the results of our analysis of the median length of negotiations and the median length of cleanup activities at SA agreement sites and similar NPL sites, which consisted of NPL sites with agreements similar to SA agreements. Appendix I includes more information on our methodology.

Median Length of Negotiations at SA Agreement Sites and Similar NPL Sites

As shown in table 3, for agreements with PRPs finalized from June 2002 through December 2012, SA agreement sites and similar NPL sites in our analysis showed mixed results in the length of time to complete negotiations, with SA agreement sites taking about as long as similar NPL sites for remedial investigation and feasibility study negotiations and less time for remedial design and remedial action negotiations.

Table 3. Median Length of Negotiations in Months for Agreements between EPA and PRPs at 66 SA Agreement Sites and 74 Similar NPL Sites for Agreements Finalized from June 2002 through December 2012

	Median length of negotiations in months		Number of negotiations for agreements between EPA and PRPs[a]	
Cleanup activities involved in negotiations[b]	SA	NPL	SA[c]	NPL
Remedial design and remedial action	7[d]	8	29	13
Remedial investigation and feasibility study	9	14	21	47

Source: GAO analysis of EPA data.

Note: Our analysis included 66 SA agreement sites that had relevant cleanup actions. Some other SA agreement sites were excluded from the analysis because of having agreements that were not relevant to our work.

[a] Sites can have more than one of each type of negotiation so the numbers of negotiations in the table do not always reflect the number of sites.

[b] EPA tracks negotiations for combined remedial investigation and feasibility study, as well as combined remedial design and remedial action in its CERCLIS database as they are defined in the *Superfund Program Implementation Manual.*

[c] In some cases at SA agreement sites, EPA did not record negotiations in CERCLIS, such as when a PRP approached EPA to begin negotiations. This occurred with two different negotiations in the SA group: one negotiation for a multisite agreement covering 11 sites and another negotiation for a multisite agreement covering 2 sites.

[d] At six SA agreement sites, EPA and the PRP negotiated a single multisite agreement that took about 3 months. However, EPA recorded a 3-month negotiation at each of the six sites in CERCLIS, which may overstate the median for the SA agreement sites in our analysis. These six sites are the only case of bundled negotiations with recorded negotiations in CERCLIS for the sites in our analysis.

Given the relatively limited number of negotiations for both NPL and SA agreement sites in our analysis and the effect of unique sites, the differences in the median length of negotiations cannot be attributed entirely to the type of approach used at each site. Unique conditions at each site have the potential to affect negotiations between EPA and the PRP beyond the cleanup approach selected.

Table 4. Median Length of Cleanup Activities for 66 SA Agreement Sites and 74 Similar NPL Sites Completed from June 2002 through December 2012

	Median number of months to complete a cleanup activity		Number of cleanup activities	
Cleanup activities[a]	SA	NPL	SA	NPL
Remedial investigation and feasibility study	69	50	14	5
Remedial design	20	22	9	20
Remedial action	31	30	6	35

Source: GAO analysis of EPA data.

Note: Our analysis included 66 SA agreement sites that had relevant cleanup actions. Some other SA agreement sites were excluded from the analysis because of having agreements that were not relevant to our work.

[a] Sites can experience more than one cleanup activity in a given phase, so the number of activities listed under each approach in the table does not always reflect the number of sites in our analysis.

Median Length of Cleanup Activities at SA Agreement Sites and Similar NPL Sites

As shown in table 4, the SA agreement sites and similar NPL sites in our analysis showed mixed results in the length of time it took to complete specific cleanup activities,[53] with SA agreement sites taking longer for remedial investigations and feasibility studies on average and about the same time for remedial designs and remedial actions on average.

Twelve of the 14 remedial investigations and feasibility studies at SA sites took longer than 50 months to complete, which is greater than the median for NPL sites in our analysis, as well as the median of 51 estimated by EPA for PRP-led remedial investigation and feasibility studies that began after June 2002. However, given the relatively small number of cleanup activities for both NPL and SA agreement sites in our analysis and differences at the site level, the differences in the median length of cleanup activities cannot be attributed entirely to the type of approach used at each site. For example, several remedial investigations and feasibility studies at SA sites took a long time to complete due to individual circumstances at the site, such as dealing with a proposal to sell on-site materials to a manufacturing company, late participation from PRPs in the process, or coordination with other cleanup efforts. SA agreement sites and NPL sites in our analysis took slightly less than 2 years on average to complete remedial designs and slightly less than 3 years on average to complete remedial actions.

End Notes

[1] CERCLA, Pub. L. No. 96-510, 94 Stat. 2767 (1980) (codified as amended at 42 U.S.C. §§ 9601- 9675 (2012)). Hereinafter, references to CERCLA sections are as amended.

[2] There is no legal requirement that EPA clean up a site on the NPL or that it do so under a particular time frame. As we have previously reported, EPA's future costs to conduct remedial construction at nonfederal NPL sites will likely exceed recent funding levels. The limited funding, coupled with increasing costs of cleanup, has forced EPA to choose between cleaning up a greater number of sites in a less time and cost-efficient manner or cleaning up fewer sites more efficiently. See GAO, *Superfund: EPA's Estimated Costs to Remediate Existing Sites Exceed Current Funding Levels, and More Sites Are Expectedto Be Added to the National Priorities List*, GAO-10-380 (Washington, D.C.: May 6, 2010).

[3] PRPs generally include current or former owners and operators of a site or the generators or transporters of the hazardous substances.

[4] Where hazardous waste sites are owned or controlled by a federal agency, that agency may also have a significant role in cleanup. Processes and provisions specific to these federal sites are

generally not discussed in this report and, according to EPA, the SA approach has not been used at a federal site.

[5] We did not assess the extent to which EPA implemented all of the IG's recommendations because it was beyond the scope of our review.

[6] For purposes of this report, deferral refers to sites where EPA elects not to use its Superfund authorities for a long-term cleanup because another program will provide oversight of the site's cleanup.

[7] Forty-seven states have followed the federal government's lead and established their own version of the Superfund program to identify and clean up sites not covered by the federal program.

[8] Operable units may address geographical portions of a site, specific site problems, or initial phases of a cleanup action.

[9] Our analysis of how SA agreement sites compare with similar NPL sites in completing the cleanup process did not include all SA agreement sites because some of these sites did not have relevant cleanup activities.

[10] These potential pathways of contamination include groundwater migration, surface water migration, soil exposure, and air migration.

[11] EPA, *Response Selection and Enforcement Approach for Superfund Alternative Sites* (Washington, D.C.: June 24, 2002); EPA, *Revised Response Selection and Settlement Approach for Superfund Alternative Sites* (Washington, D.C.: June 17, 2004); and EPA, *Updated Superfund Response and Settlement Approach for Sites Using the Superfund Alternative Approach (SAA)* (Washington, D.C.: Sept. 28, 2012).

[12] For non-NPL sites, EPA does not have the option to perform a long-term cleanup itself and seek reimbursement from PRPs; however, EPA may enter into settlements with PRPs where applicable conditions for enforcement actions are met.

[13] The Government Performance and Results Act of 1993, Pub. L. No. 103-62, 107 Stat. 285 (1993), amended by the Government Performance and Results Act Modernization Act of 2010, Pub. L. No. 111-352, 124 Stat. 3866 (2011).

[14] CERCLA § 122(a), 42 U.S.C. § 9622(a) (2012).

[15] CERCLA § 122(d)(1)(A), 42 U.S.C. § 9622(d)(1)(A) (2012). EPA tracks separately the duration of negotiations for (1) remedial investigation and feasibility study agreements and (2) remedial design and remedial action agreements.

[16] CERCLA § 106(a), 42 U.S.C. § 9606(a) (2012).

[17] Pub. L. No. 94–580, 90 Stat. 2795 (1976) (codified as amended at 42 U.S.C. §§6901- 6992k (2012)). Although RCRA amended the Solid Waste Disposal Act, Pub. L. No. 89– 272, Title II, 79 Stat. 997 (1965), the amended law is nonetheless sometimes referred to as RCRA, a convention we follow here. Subtitle C of RCRA, 42 U.S.C. ch. 82, subch. III (§§ 6921-6939f), governs hazardous waste management. Hereinafter, references are to RCRA as amended.

[18] See, e.g., 10 U.S.C. § 2701 (2012), giving DOD responsibility for environmental restoration at its facilities, among other things. See also Executive Order 12580, Superfund Implementation, 52 Fed. Reg. 2923 (Jan. 23, 1987). The executive order was issued in 1987 to respond to the Superfund Amendments and Reauthorization Act and delegates to EPA certain regulatory authorities that the statute assigns to the President, while delegating to the Departments of Defense and Energy authority for removal and remedial actions at their facilities, subject to certain provisions of CERCLA. This executive order generally gives other federal agencies authority for cleanups at non-NPL listed sites.

[19] Federal facility sites on the NPL are subject to certain additional requirements, see 42 U.S.C. § 9620(e) (2013).

[20] See 48 Fed. Reg. 40,658 (Sept. 8, 1983) (discussing original deferral policy), 60 Fed. Reg. 14,641 (Mar. 20, 1995) (providing explanation of deferral policy and revisions to RCRA policy), 54 Fed. Reg. 41,000 (Oct. 4, 1989) (discussing EPA development of its approach to RCRA deferral), 53 Fed. Reg. 23,978 (June 24, 1988) (discussing revisions to RCRA deferral). See also EPA Memorandum, Coordination between RCRA Corrective Action and Closure and CERCLA Site Activities (1996); and EPA, "NPL Deletion/Deferral Policy and RCRA Subtitle C Corrective Action," EPA: 540-R-95-002g (1995).

[21] We identified the approximately 3,400 sites based on whether the site was undergoing long-term cleanup. According to EPA officials, sites under each of the long-term cleanup approaches would have a Hazard Ranking System score of at least 28.50, otherwise the site would have been classified as "no further remedial action planned." These sites do not include sites that EPA has archived in CERCLIS.

[22] States also oversee cleanup at an undetermined additional number of sites that may be eligible for the NPL but are not listed in CERCLIS. States are not obligated to report all potentially eligible sites to the Superfund program, and environmental officials in several states confirmed that they have conducted or overseen cleanups at sites not listed in CERCLIS that may have been eligible for the NPL.

[23] NPL sites do not include sites that have been deleted from the NPL.

[24] EPA has had SA agreements at 70 sites in total; 3 of these sites have switched from the SA approach to the NPL approach.

[25] EPA officials said the Superfund program had deferred 10 sites to the RCRA program in the last 5 years, and that deferrals from the RCRA program to the Superfund program probably have been more common recently, because of business bankruptcies. Active hazardous waste treatment, storage, or disposal facilities at which contamination has occurred may be eligible to be cleaned up as corrective actions in the RCRA program. These cleanups are generally the responsibility of the party owning or operating the permitted facility. If the party goes bankrupt, the party may be unable to complete the RCRA corrective action.

[26] EPA Office of Solid Waste and Emergency Response, *Guidance on Deferral of NPL Listing Determinations While States Oversee Response Actions*, EPA1540/F-95/002 (Washington, D.C.: May 1995), 4.

[27] CERCLA also establishes requirements related to formal state deferrals; see 42 U.S.C. § 9605(h) (2012).

[28] All six of these regions had at least one site identified as an OCA deferral to private parties.

[29] The completion of cleanup date at an OCA deferral site is the date of the determination that cleanup was successfully completed, that cleanup was not necessary, or that the other entity will not complete cleanup and the site will be referred back to the Superfund program.

[30] Specifically, CERCLA established a trust fund from which EPA receives annual appropriations for Superfund program activities. Superfund trust funds are available for long-term cleanups only at sites on the NPL. EPA may use these funds for investigation and removal actions at any nonfederal site. EPA also can seek reimbursement from the PRPs after incurring these costs.

[31] NRD claims are made for injury to, destruction of, or loss of natural resources. CERCLA defines natural resources broadly to include land, fish, wildlife, groundwater, and other resources belonging to or managed by federal or other governmental entities.

[32] Under certain circumstances—for example, if EPA determines that a PRP that has entered into an SA agreement for a site is not adequately fulfilling the requirements of the agreement—EPA may decide to list the site on the NPL.

[33] EPA IG, *EPA Needs to Take More Action in Implementing Alternative Approaches to Superfund Cleanups,* 2007-P-00026 (Washington, D.C.: June 6, 2007).

[34] EPA Office of Enforcement and Compliance Assurance, "Certification to Close OIG Audit Report, 'EPA Needs to Take More Action in Implementing Alternative Approaches toSuperfund Cleanups'" (Washington, D.C.: Aug. 30, 2010).

[35] EPA IG, *EPA Needs to Take More Action,* 6.

[36] GAO, *Standards for Internal Control in the Federal Government,* GAO/AIMD-00-21.3.1 (Washington, D.C.: November 1999).

[37] See, for example, EPA, *The Office of Solid Waste and Emergency Response, Fiscal Year 2010 End of the Year Report* (Washington, D.C.). Available at http://www.epa.gov/superfund/accomplishments.htm. Accessed March 22, 2013.

[38] EPA IG, *EPA Needs to Take More Action,* 10.

[39] EPA Office of Enforcement and Compliance Assurance, "Certification to Close OIG Audit Report," 2.

[40] Region 4 prepares a draft Hazard Ranking System documentation record for SA agreement sites because the region wants the site to be ready for listing on the NPL, if necessary. In addition, in some cases, sites were already proposed to the NPL when the SA agreement was entered, negating any savings from avoiding this step.

[41] See, for example, EPA, "Superfund Alternative Approach Baseline Assessment" (Washington, D.C.: Apr. 14, 2011); GAO, *Superfund: Greater EPA Enforcement and Reporting Are Needed to Enhance Cleanup at DOD Sites,* GAO-09-278 (Washington, D.C.: Mar. 13, 2009); Industrial Economics Incorporated, *Effectiveness Assessment of the Region 4 Superfund Alternative Approach,* a report prepared for the EPA, November 2010.

[42] Industrial Economics Incorporated, *Effectiveness Assessment,* ES-7.

[43] Industrial Economics Incorporated, *Effectiveness Assessment,* ES-8.

[44] Industrial Economics Incorporated, *Effectiveness Assessment,* ES-9.

[45] Multiple cleanup activities can occur within a given phase at the same or different operable units at one site. Completion of one cleanup activity, such as a remedial investigation and feasibility study, does not necessarily mean all work in that phase has been completed. Our analysis looks at individual activities within given phases.

[46] We only reported information from CERCLIS's active inventory of sites, as opposed to the archived inventory. EPA archives sites when no further interest exists at the site under the federal Superfund program based on available information.

[47] We excluded four SA agreement sites that did not have relevant agreements involving a PRP-led combined remedial investigation and feasibility study, remedial design, or remedial action. First, one agreement involved EPA's use of a special account where the PRP funded the initial remedial investigation work based on the understanding that the PRP would conduct remedial action. Second, another site's SA agreement involved a PRP-led removal action and no SA agreement for remedial investigation and feasibility study, remedial design, or remedial action. Third, one SA agreement site had no legal action data. Finally, at one SA agreement site, the PRP began conducting cleanup activity before establishing an SA agreement with EPA.

[48] We analyzed administrative orders on consent and consent decrees separately because consent decrees are used for remedial actions which can have higher costs than the activities under an administrative order on consent.

[49] Consent decrees, which are used for remedial actions, must be approved by a court. Due to this requirement, there can be delays between completion of negotiations and the decree becoming official, so we included agreements with completed consent decrees within 2 years of the end of the negotiation.

[50] The one exception involved remedial actions, with NPL comparison sites taking 20 months longer, on average, to complete remedial actions than their SA agreement site counterparts in our regional analysis restricted to regions 4 and 5. However, this comparison involved only a small number of NPL sites since Regions 4 and 5 only contain around one-third of the complete set of NPL comparison sites used in our analysis.

[51] Generally, a site is considered to be a megasite if the combined extramural, actual and planned, removal, and remedial action costs incurred by Superfund or by PRPs are greater than $50 million. The megasite designation may be applied to any federal or non- federal facility NPL or non-NPL site.

[52] While the duration of cleanup activities was somewhat sensitive to the distribution of megasites, it was not always in the expected direction. For example, SA agreement sites in which megasites were excluded took longer on average to complete remedial action activities than the entire SA group. In contrast, NPL comparison sites in which megasites were not included took less time, on average, to complete remedial investigation and feasibility study activities than the entire set of NPL comparison sites.

[53] Multiple cleanup activities can occur within a given phase at the same or different operable units at one site. Completion of a cleanup activity, such as a remedial action, does not necessarily mean the site has completed the entire phase.

INDEX

A

access, 10, 62, 67
accounting, 39, 66, 70
adverse weather, 29
age, 11, 35, 36, 37, 38
agencies, 6, 32, 40, 49, 56, 57, 58, 61, 69, 70, 84, 85, 89
Air Force, 61
Alaska, 37
American Recovery and Reinvestment Act, 3, 13, 32
American Recovery and Reinvestment Act of 2009, 3, 13, 32
appropriations, vii, 1, 2, 4, 5, 9, 13, 14, 19, 20, 22, 23, 26, 28, 31, 32, 41, 70, 78, 90
arsenic, 4
assessment, 6, 52, 69, 70, 72, 80
assets, 26, 42
audit, 6, 34, 51, 83
authorities, 49, 57, 60, 65, 89
authority, 9, 33, 56, 57, 65, 66, 89

B

bankruptcies, 25, 90
base, 42
batteries, 7
breakdown, 61, 84

C

caliber, 79
cancer, 4
carbon, 41
carbon atoms, 41
cash, 10
category a, 32
Census, 11, 12, 34, 38, 41
CERCLA, 3, 9, 10, 41, 47, 51, 56, 57, 65, 67, 79, 88, 89, 90
chemical, 4, 41
chemicals, 7, 9
chlorinated hydrocarbons, 41
chlorine, 41
citizens, 6, 49
cleaning, 4, 9, 26, 41, 42, 47, 88
cleanup, vii, 1, 2, 4, 5, 7, 8, 9, 10, 13, 14, 19, 20, 21, 22, 23, 25, 26, 29, 32, 33, 40, 41, 42, 45, 46, 47, 48, 49, 50, 51, 52, 53, 55, 56, 57, 58, 59, 60, 61, 62, 63, 64, 65, 66, 67, 68, 70, 71, 72, 73, 75, 76, 77, 78, 79, 80, 81, 82, 83, 84, 86, 87, 88, 89, 90, 91, 92
coding, 77
color, iv
commercial, 57
communities, 53, 66, 71, 72, 73
community, 17, 66, 67, 73
complexity, 11, 28, 83

compliance, 69
compounds, 41
Comprehensive Environmental Response, Compensation, and Liability Act, 3, 47, 48
Comprehensive Environmental Response, Compensation, and Liability Act (CERCLA), 3, 48
conference, 50, 80
Congress, iv, 56, 70, 78
consent, 56, 81, 82, 91, 92
construction, vii, 2, 4, 5, 7, 8, 9, 14, 15, 18, 19, 24, 29, 30, 32, 41, 42, 43, 55, 74, 75, 76, 82, 88
contaminant, 16
contaminated sites, vii, 1, 4, 10
contaminated soil, 4
contaminated water, 58
contamination, vii, 1, 4, 7, 17, 24, 41, 42, 48, 49, 50, 52, 53, 55, 60, 61, 67, 72, 77, 78, 80, 89, 90
convention, 89
cooperative agreements, 51
coordination, 75, 88
cost, 10, 16, 17, 20, 32, 41, 42, 55, 71, 88
counsel, 59
court approval, 56
covering, 75, 87
crude oil, 9

D

damages, iv, 48, 67
danger, 57
database, 5, 10, 32, 47, 49, 50, 58, 68, 78, 79, 80, 83, 87
decision-making process, 59
decontamination, 57
defects, 4
delegates, 89
Department of Defense, 57
Department of Justice, 56
destruction, 90
directives, 69
distribution, 82, 83, 92

District of Columbia, 12, 34, 37, 38, 41
draft, 31, 65, 69, 79, 91

E

e-mail, 31, 64
emergency, 49
emergency response, 49
endangered, 16
endangered species, 16
enforcement, 14, 56, 79, 89
environment, 4, 6, 7, 8, 9, 16, 48, 49, 51, 52, 53, 55, 57
Environmental Protection Agency, vii, 2, 3, 4, 32, 46, 47, 48, 79
environmental regulations, 57, 58
environmental threats, 25
EPA, i, iii, v, vii, 1, 2, 3, 4, 5, 6, 7, 8, 9, 10, 11, 12, 13, 14, 15, 16, 17, 18, 19, 20, 21, 22, 23, 24, 25, 26, 27, 28, 29, 30, 31, 32, 33, 34, 38, 39, 40, 41, 42, 43, 45, 46, 47, 48, 49, 50, 51, 52, 53, 54, 55, 56, 57, 58, 59, 60, 61, 62, 63, 64, 65, 66, 67, 68, 69, 70, 71, 72, 73, 74, 75, 76, 77, 78, 79, 80, 81, 82, 83, 84, 85, 86, 87, 88, 89, 90, 91
equipment, 7, 16
evidence, 6, 34, 51, 80, 83
Executive Order, 89
expenditures, vii, 1, 2, 5, 13, 14, 19, 20, 21, 23, 32, 41, 42
exposure, 4, 70, 89

F

fear, 73
federal agency, 57, 88
federal facilities, 4, 58
federal government, 48, 89
Federal Government, 91
federal law, 50
Federal Register, 6, 9, 26, 33, 53, 71, 73
financial, 5, 26, 32, 42, 62, 71
financial data, 32, 42

financial resources, 62
financial system, 5, 32
fiscal year, vii, 1, 2, 4, 5, 9, 10, 11, 12, 13,
    14, 17, 18, 19, 20, 21, 22, 23, 24, 25, 26,
    27, 28, 29, 30, 31, 32, 33, 34, 38, 39, 41,
    42, 43, 79
fish, 90
funding, 1, 5, 13, 16, 17, 18, 19, 25, 32, 41,
    42, 47, 59, 66, 88
funds, 2, 9, 10, 13, 14, 17, 18, 19, 20, 22,
    23, 28, 29, 32, 33, 41, 42, 66, 67, 72, 90

G

GAO, 1, 2, 8, 12, 14, 15, 16, 18, 20, 21, 22,
    23, 26, 27, 28, 30, 31, 33, 38, 39, 40, 41,
    42, 45, 46, 47, 52, 54, 55, 60, 74, 76, 84,
    85, 86, 87, 88, 91
Georgia, 36
GPRA, 47, 56, 70, 78, 79
grants, 66
groundwater, 7, 17, 40, 41, 42, 55, 89, 90
Guam, 12, 34, 38
guidance, 34, 41, 46, 48, 50, 53, 56, 57, 58,
    59, 63, 64, 65, 66, 67, 68, 69, 77, 78, 79,
    80, 81

H

habitat, 16
Hawaii, 37
hazardous materials, 4, 57
hazardous substance, 4, 6, 9, 10, 41, 52, 57,
    88
Hazardous Substance Superfund Trust
    Fund, 3, 9
hazardous substances, 4, 6, 9, 10, 41, 52, 57,
    88
hazardous waste, 4, 6, 42, 45, 48, 49, 50, 51,
    56, 58, 80, 82, 88, 89, 90
health, 4, 6, 7, 17, 52, 57, 60
health problems, 4
House, 48
House of Representatives, 48

housing, 43
human, 4, 6, 7, 8, 9, 16, 25, 48, 49, 51, 52,
    53, 55, 70
human exposure, 70
human health, 4, 6, 7, 8, 9, 16, 25, 48, 49,
    51, 52, 53, 55
hybrid, 65
hydrogen, 41

I

identification, 63
identity, 69
improvements, 47
individuals, 73
infertility, 4
infrastructure, 16
injury, iv, 90
internal controls, 70
investments, 41
Iowa, 36

J

justification, 69

L

landfills, 7
laws, 46
lead, 4, 17, 51, 76, 89
litigation, 71
local government, 41, 53, 72
Louisiana, 36

M

majority, 2, 13, 20, 22, 41, 58, 61, 82
management, 7, 16, 40, 41, 47, 58, 64, 69,
    78, 89
manufacturing, 7, 25, 32, 75, 88
Maryland, 36
materials, 75, 88

matter, iv, 6
median, 13, 21, 23, 25, 26, 29, 30, 32, 33,
    42, 82, 86, 87, 88
metals, 7
methodology, 6, 12, 31, 34, 38, 51, 74, 79,
    86
Mexico, 36
migration, 89
military, 61
Missouri, 35
Montana, 36
motivation, 72

N

National Academy of Sciences, 33
National Priorities List, i, iii, v, vii, 1, 2, 3,
    4, 8, 12, 16, 18, 20, 21, 22, 23, 26, 27,
    28, 30, 31, 32, 35, 38, 39, 40, 41, 42, 45,
    46, 47, 48, 79, 88
natural resources, 90
negotiating, 56
negotiation, 75, 82, 83, 86, 87, 92
NPL, vii, 1, 2, 3, 4, 5, 6, 7, 9, 11, 13, 14, 15,
    19, 20, 21, 24, 25, 26, 27, 28, 29, 30, 32,
    33, 34, 35, 36, 37, 38, 39, 40, 41, 42, 43,
    45, 46, 47, 48, 49, 50, 51, 53, 54, 56, 57,
    58, 59, 60, 61, 62, 65, 66, 67, 68, 69, 70,
    71, 72, 73, 75, 76, 77, 78, 79, 80, 81, 82,
    83, 84, 85, 86, 87, 88, 89, 90, 91, 92
NRC, 47, 57, 62, 63, 84, 85
Nuclear Regulatory Commission, 47, 57
Nuclear Regulatory Commission (NRC), 57

O

officials, 2, 4, 5, 6, 10, 11, 13, 16, 17, 18,
    19, 20, 22, 23, 24, 25, 26, 28, 29, 30, 32,
    33, 34, 41, 42, 43, 46, 47, 50, 51, 52, 53,
    55, 56, 57, 59, 60, 61, 62, 63, 64, 65, 67,
    68, 69, 70, 71, 72, 73, 75, 76, 77, 80, 81,
    82, 83, 90
oil, 7
Oklahoma, 22, 37

operations, 7, 40, 51
opportunities, 73
organic chemicals, 41
outreach, 26, 73
overlap, 8, 55
oversight, 41, 45, 46, 49, 50, 51, 53, 56, 57,
    58, 63, 64, 67, 72, 77, 78, 80, 89

P

pathway, 41
pathways, 52, 89
payroll, 14
permission, iv
policy, 6, 16, 25, 59, 60, 72, 90
pollutants, 4, 6
polychlorinated biphenyl, 4
polychlorinated biphenyls, 4
population, 11, 16, 20, 33, 34, 35, 36, 37,
    38, 43
population size, 16
preparation, iv, 71
preservation, 7
President, 14, 42, 89
private party, 56, 58
probability, 43
project, vii, 2, 4, 5, 8, 9, 17, 24, 28, 29, 32,
    41, 42, 43, 71
protection, 8, 55
Puerto Rico, 12, 34, 38

Q

qualifications, 72

R

Radiation, 22
reactions, 73
recognized tribe, 57
recommendations, iv, 2, 46, 48, 79, 80, 89
recovery, 7, 41
recycling, 7, 25, 42, 57
redevelopment, 71, 72

regulations, 46, 50, 80
reimburse, 10, 19
reliability, 5, 34, 43, 50, 83
remedial actions, 4, 47, 66, 74, 75, 82, 88, 89, 91, 92
remediation, 42, 73
removal action, 4, 49, 53, 90, 91
removal actions, 4, 49, 90
requirement, 41, 48, 73, 88, 92
requirements, 46, 53, 56, 73, 90, 91
reserves, 61
resources, 10, 26, 59, 62, 70, 73, 90
response, 7, 8, 9, 10, 15, 26, 55, 56, 60, 66, 84, 85
restoration, 7, 89
revenue, 9
rights, iv
risk, 4, 9, 16
risks, 7, 16, 48, 55

S

safety, 60
sampling error, 43
savings, 91
scope, 5, 31, 32, 51, 74, 79, 89
sediment, 4, 20
sediments, 7, 40
Senate, 3
sensitivity, 83
September 11, 31
services, iv, 10, 41
settlements, 10, 89
signs, 16
site cleanup, 9, 10, 25, 53, 77
software, 11
South Dakota, 37
stability, 16
staffing, 28, 29
stakeholders, 33, 59
state, 5, 6, 7, 10, 11, 25, 26, 29, 32, 33, 35, 36, 37, 38, 41, 49, 50, 51, 56, 57, 58, 59, 61, 62, 63, 64, 67, 69, 71, 72, 80, 84, 85, 90

states, 2, 10, 12, 13, 20, 25, 34, 38, 41, 42, 46, 50, 51, 53, 56, 57, 59, 61, 62, 63, 64, 66, 67, 71, 72, 80, 84, 85, 89, 90
statute of limitations, 67
statutes, 80
stigma, 48, 72, 73
storage, 56, 90
subgroups, 43
Superfund, i, iii, v, vii, 1, 2, 3, 4, 5, 6, 8, 9, 10, 13, 14, 15, 19, 25, 28, 29, 31, 32, 33, 40, 41, 42, 45, 46, 47, 48, 49, 50, 51, 53, 55, 56, 58, 60, 61, 62, 63, 64, 65, 68, 69, 70, 77, 78, 79, 80, 83, 84, 85, 87, 88, 89, 90, 91, 92

T

taxes, 9
teams, 59
technical assistance, 66, 67, 72
technical comments, 31, 79
techniques, 66
threats, 2, 4, 7, 25, 49, 55
time frame, 19, 25, 29, 88
Title I, 89
Title II, 89
toxicity, 16
tracks, 58, 68, 70, 87, 89
transportation, 56
treatment, 7, 56, 90
trust fund, 90
Trust Fund, 3, 9, 10
trust funds, 90

U

United States, v, 1, 3, 40, 45
universe, 51, 69, 83

V

variables, 82

W

Washington, 35, 41, 42, 88, 89, 90, 91
waste, 7, 23, 25, 48, 51, 52, 57
waste disposal, 7

waste management, 23, 25
water, 89
websites, 63
wildlife, 90
Wisconsin, 35
wood, 7